Bibliotheca Indo-Buddhica Series No. 107

Inference and Fallacies Discussed in Ancient Indian Logic

(with special reference to Nyāya and Buddhism)

Pradeep P. Gokhale

Sri Satguru Publications
A Division of
Indian Books Centre
Delhi - India

Published by :
Sri Satguru Publications
Indological and Oriental Publishers
A Division of
Indian Books Centre
40/5, Shakti Nagar,
Delhi-110007
(INDIA)

First Edition: Delhi 1992

ISBN 81-7030-319-2

Printed in India

S. K. Mehra at Mehra Offset Press, 1976, Chandni Mahal Darya Ganj, New Delhi-110002

Foreword

When I was doing philosophy as an undergraduate the social and philosophical atmosphere of India was conditioned by Anglo-Saxon and European thought. The histories of Indian philosophy written by Dasgupta and Radhakrishnan were published only in the third decade of 20th Century. Of course it was not the case that scholars had paid no heed to Indian philosophy. From Jocobi onwords, the European scholars like Max Muller and A.B. Keith had taken to the study of Indian philosophy. And even Indian scholars like Vidyabhushana, Ganganath Jha, Kuppuswami Shastri and several others did write on Indian Philosophy. But such studies had remained only the private property of the orientalists. What was true of Indian philosophy was even in greater measure true of the study of Indian logic. Some scholars like Stcherbatsky did write on Buddhist logic and Vidyabhushan's *History of Indian Logic* would ever remain a guiding star for the explorers of Indian logic, but by and large Indian logic like Indian philosophy was the monopoly of Orientalists and Indian logic, particularly of the Navyanyāya variety was the monopoly of the scholars in Bengal. In fact Indian scholars should be grateful to pandits from Bengal for it is they who preserved the traditional study of Navyanyāya texts in well knit form. Not that there were no oases of Nyāya scholarship in other places; Varanasi in U.P. was an abode of several Nyāya scholars. Like Varanasi, Triupati in South India and some places in Bihar (perhaps Mithila) had given shelter to Navyanyāya scholars. Raja Rama Verma of the ruling class of Kochin was regarded as one of the greatest Nyāya scholars in South India. Already a reference has been made to Kuppuswami Shastri. In Maharashtra too, Shri Athley's work on Tarka Sangraha attained a classical status. But perhaps Prof. Randle of Allahabad University was amongst the first few who took to the study of Indian logic in contemporary idiom. This was followed by Ingalls who worked under Kalipada Tarkacharya and Prof. Shaileshwar Sen. This process was continued by Father Bochenski, J.F. Stall, Karl Potter; Schayer and Kunst of the school of the Polish logician, Lukasiewicz.

While European and American scholars has started studying logic of Indian variety in the new idiom, Indians were still unaware of the developments in western logic itself. Several European and American scholars had started interpreting logic in mathematical terms. This new logic was known as Mathematical logic or Symbolic Logic. The leaders of this movement were Russell in England, Hilbert in Germany, Brouwer in Netherlands, Lasniewski and Lukasiewicz in Poland, Wittgenstein, philosophers of Viennese circle and Pierce of America also made important contributions in this area. Mathematical logic was also being done by Chinese scholars like Hao Wong. Developments in the area of Mathematics by Peano in Italy, Sheffer in Norway, Poincare in France and Georg Cantor in Germany had helped the growth of this logic.

When in Oxford I had studied some of these developments under Prof. Thomson, Hao Wong, Kneale, Lemmon and Dummet. In fact Prof. Dummet who was lecturing then on set theory had asked me whether Indian logicians had made any development in this direction.' But at that time I was quite unaware of Indian logic itself. While in England Prof. Dorothy Emmet had invited me to Manchester and had asked me to give a seminar on Indian theories of meaning. In fact she impressed on me the necessity of studying Indian Philosophy and logic in the light of the recent developments in Europe and America.

But when I came to India I found that the study of Indian logic was still discouraged in the Indian universities. However, my joining S.V.University was a great boon to me. First, Tirupati was a Centre of traditional study for Indian logic of Nyāya variety and secondly I found a colleague in Dr. S. Sarma of Sanskrit Department who was trained in Nyāya. It is here that I took lessons in Navyanyāya under Shri N.S. Rajgopal Tatachari. Earlier I had read *Tarkasangraha* with C. Chandrodaya Bhattacharya and had taken some lessons in Navyanyāya logic under Shri Atmaram Shastri Jere. Thus my training in Symbolic logic in Oxford and Navyanyāya logic in Amalner and Tirupati made me think that the tradition of Indian logic should be kept alive and that as far as possible it should be understood in the new idiom of Western logic.

Perhaps during this period several teachers in Indian universities must have thought in similar terms with the result that in the universities of Delhi, Allahabad, Calcutta, Jadhavpur and Jaipur some people had already started doing some homework in Indian logic. This was the time when Dr. Narayanshastri Dravid was appointed in Nagpur University. He was expert in Nyāya and had studied Diṅnāga and Dharmakīrti. At some stage I had also thought that the works of Indian mathematicians and grammarians should be studied for enriching the theories of Indian logic.

Perhaps my study of Indian logic indirectly shaped the thinking of some of my students who did much greater progress than me in this area. Prof. Dr. C. Ramayya and Prof. Dr. Virayya of S.V. University, Dr. Bankelal Sharma of Punjab University, Dr. V.K. Bharadwaja of Delhi and Dr. S.E. Bhelke, Dr. Pradeep Gokhale and Dr.Mrs.Ketkar of Poona University did some important work in the area of Indian logic. Studying Indian logic became more or less a movement and what I did in Poona University was done with the some vigour in Jadhavpur, Jaipur, Delhi and Allahabad Universities.

With the blessings of Wrangler Mahajani who was then the Vice-Chancellor of Poona University, I could get Nyaya-Sarvabhauma Sitaramshastri Kurundkar appointed as Nyaya Pandit in Poona University. Even scholars from other countries, particularly from Japan studied Indian logic in our department under Shri Kurundkar Shastri, and later under Shri Sudhakar Dixit and Dr. Baliram Shukla.

The present work of Dr. Pradeep Gokhale is at least in some sense the result of change of atmosphere in this country and the change of outlook in Poona University Department of Philosophy.

It is an accident that I met Dr. Pradeep Gokhale during my tenure as Rector of the University. He was well-versed in Sanskrit and perhaps was thinking of doing M.A. in Statistics or Sanskrit. I found that he was a good poet and very soon I discovered that he had a potentiality to be a good philosopher and logician. Fortunately he was attracted towards philosophy, left his study of Sanskrit and Statistics and joined philosophy. When he decided to do research in philosophy, he choose for his research the topic of fallacies in Indian logic and requested me to guide him. The work in this area was still not attempted. Prof. Rasik Bihari Joshi of Delhi University Department of Sanskrit had written a small book on fallacies but it was more in a traditional line. Dr. Gokhale's work, I thought, would cut the ice. I am aware that simultaneously with Dr. Gokhale a good research work on fallacies was produced from Benares Hindu University. But I believe that Dr. Pradeep Gokhale's work is unique in this area.

Dr. Pradeep Gokhale's present book which in some sense is an adaptation of his Ph.D. work on fallacies would be very useful to the students and researchers of Nyāya, Buddhist and Jaina logics. It is mainly an exercise in the history of logico-philosophical ideas, especially the concepts of *anumāna* and *hetvābhāsa. Anumāna and hetvābhāsa* are the central concepts of Indian logic. We sometimes treat these concepts as if they remained unchanged throughout the history of Indian logic and forget that they did undergo radical change. One of the tasks of Dr. Gokhale in this work is to sketch out the way in which the concepts of *anumāna and hetvābhasā* changed and developed, individually and mutually, at the hands of ancient Nyāya and Buddhist logicians. He claims that the early Nyāya concept of *anumāna* was that of analogical reasoning. That *anumāna* should be valid and sound and not just a strong analogical reasoning was the concept of *anumāna* newly developed by Dharmakīrti. Dr. Gokhale also shows how the Nyāya logicians after Dharmakīrti tried to unite these two concepts together. In fact Dharmakīrti plays a pivotal role in the history of Indian logic. Clear understanding of this role is essential for tracing the development of Indian logic. Dr. Gokhale's account of this role is worth critical consideration.

Another important contribution of Dr. Gokhale in this work is the comparative survey of Indian and Western accounts of fallacies. Western logicians attach central importance to formal fallacies and only secondary one to material fallacies. Hetvābhāsas which are central in Indian logic are however material fallacies. Dr. Gokhale claims that *sādhanābhāsa* rather than *hetvābhāsa* is the proper Indian analogue of 'fallacy' and that *sādhanābhāsa* covers material as well as formal fallacies. In addition to *sādhanābhāsas* (which include *hetvābhāsas* proper, Dr. Gokhale also discusses *chala, jāti, nigrahasthāna, nayābhāsa,* fallacies of definition and *prasaṅga* fallacies in detail and their place in the Indian theory of fallacy. This too might prove a unique contribution of this work.

The work is full of observations, interpretations and arguments of historical and philosophical logical character. I am sure this work will be appreciated by logicians in India and outside. I need not write anything more about the book, for the book itself would speak for its worth.

I am happy that Dr. Gokhale asked me a to write the foreword to this important book. I wish that Dr. Gokhale writes many more books of this kind and makes Indian logic and philosophy richer.

S.S. Barlingay

Contents

Preface

A few treatises which seriously discuss Indian account of inference and fallacies are now readily available. Why then this new addition? The only justification that I have is this : I have tried in this book to develop an approach to the problem which has been more or less undeveloped hitherto.

Thus I have attempted

1. to outline the development of the ancient Nyāya and Buddhist conceptions of inference and fallacy.

2. to discuss in detail the concept of fallacy and the classification of fallacies in the light of ancient Indian account of the same.

3. to compare ancient Indian awareness of inference and fallacies with the western awareness of the same.

In traditional approach to ancient Nyāya logic we find a tendency to consider 'Nyāya view' to be a fixed set of doctrines and principles whch did not undergo any change or development since Akṣapāda to Udayana. It is held that what was implicitly or explicitly there in the Nyāya-aphorisms was made explicit or more explicit and was applied without any modifications to refute the doctrines of the rival schools, by the commentators of the *Nyāyasūtra*. So if by development we understand the modifications of earlier principles and doctrines and discovery of new ones, the ancient Nyāya system did not really develop. Such an appraoch is rather unhistorical and unfair too. Here I have approached ancient Nyāya with a contrary belief. I believe that the Nyāya system must have developed through the works of the commentators of *Nyāyasūtra*. This development of Nyāya must have been partly due to its interaction with other schools of logic, especially Buddhism. When I started reading the ancient commentaries of *Nyāyasūtra* in the light of this belief the commentators started telling me more about their own contributions and importations (from other schools) than about the *sūtras* they were commenting upon. In fact Indian Philosophy in general and Indian logic in particular has grown in this peculiar way. Commentators have contributed to the development of the systems when they themselves humbly submitted that they only tried to elucidate and defend the original work as it was. Accordingly in the first six chapters of this treatise I have tried to outline the development of Ancient Nyāya and Buddhist conceptions of inference and fallacy, as it was effected by the original authors and commentators of the two schools. I have also tried to show how the interaction between these two schools must have contributed to their development.

Many books dealing with Indian logic or some aspects thereof concentrate on the classical indian concept of *anumāna* quite thoroughly but the topic of *hetvābhāsa* or *sādhanābhāsa* remains comparatively neglected. But possibly a clearer understanding of what is a sound inference or what is *anumāna-pramāṇa* can be had if we have a sufficiently clear understanding of what it is not (or what it is opposed to). Keeping this in mind I have, in this book, tried to give equal weightage to the topics of inference and fallacies. Such an attempt has helped me to see how the ancient Indian conceptions of inference and fallacy developed reciprocally and mutually rather than independently and separately. However, the variety of ways in which inference can be unsound and therefore fallacious is far greater than the variety of ways in which it can be sound. The reader will therefore find more pages of this book devoted to the discussion of fallacies than those to that of inference.

After having looked closely into the different kinds and varieties of formal and informal fallacies discussed in our ancient philosophical literature, I have tried to search for a possible Indian analogue of the western concept of fallacy. I have tried to show that the term *sādhanābhāsa* as it was used by Diṅnāga and Dharmakīrti, rather than the term *hetvābhāsa* is the proper Indian analogue for the western term fallacy. I have also argued that apart from *pakṣābhāsa*, *hetvābhāsa* and *dṛṣṭāntābhāsa* which the Buddhist tradition subsumes under *sādhanābhāsa*, some other varieties of fallacy to the extent to which they are properly so, viz. *chala*, *jāti*, *nigrahasthāna*, *prasaṅga*, *nayābhāsa* and fallacies of definition can also be brought under the common head *sādhanābhāsa*.

While comparing Indian awareness of inference and fallacies with the western one, I have shown that Indian concept of *anumāna* is different from the western concept of inference in some very important respects. I have also tried to show that the attempt of some orientalists to identify the cases of *hetvābhāsa* with the cases of Aristotalian fallacies of syllogism is misleading. I have also pointed out that Dharmakīrti has tackled the questions of formal validity and formal fallacies in their proper perspectives. This gives a different dimension to Indian logic which many orientalists ignore.

Some modern scholars of Indian logic have attempted to present the ideas of Indian logicians in the language of modern (symbolic) logic. This way of presentation, if done properly, has some advantages. For instance it makes it possible to have more lucid and sharper understanding of the laws of logic stipulated or discoverd by Indian logicians. Some books which attempt this with reference to Navya- Nyāya frame of inference are available to us. In this book (in Appendix III) I too, have attempted it with reference to the forms of *parārthānumāna* accepted by Vatsyāyana, Diṅnāga and Dharmakīrti.

Now some formal details: I have divided some chapters into parts and all the chapters into sections. Generally the chapters concerned with the account of inference and fallacy, are divided into two parts, the first of which deals with 'inference' and the second with 'fallacy'.

While mentioning their names, I have distinguished between classical authors on fallacies etc., and modern scholars. I have mentioned the names of the latter in all capitals and those of the former in the usual way.

While giving the history of the concepts of inference and fallacy from Caraka to Vācaspatimiśra and while giving the Nyāya-Buddhist debate on these concepts, which is the subject matter of the first six chapters, I did not concern myself with the problems regarding the exact dates of these authors. Regarding these matters I depended on the orientalist scholars. Of course, regarding the order of these authors as to who came after whom, there does not seem to be any controversy except in the case of Jayanta, Bhāsarvajña and Vācaspatimiśra. In such controversial matters I have followed POTTER.

While writing the book I had to face the problem of language. I came across many Sanskrit terms, for which I could not find apt synonyms in English. Sometimes I had to avoid even the popular synonyms because I thought them misleading. For instance I thought that 'reason' or 'probans' would not be an apt synonym for '*hetu*'. Though the term *hetu* sometimes means 'reason', generally *hetu* is taken to mean some ontological object which functions as the 'signifying analogy' or at least 'signifying object'. Thus *hetu* is, many a time, ontological and this sense of *hetu* cannot be conveyed by the words like reason or probans. Secondly reason or probans are wider concepts in the sense that even a statement of *vyāpti* in an inference can be called the reason or probans as far as it functions as the justification for proving the probandum. But the statement of *vyāpti* or *vyāpti* itself cannot be called *hetu*. Similar difficulties arose in my mind regarding the translations of *vyāpti, parāmarśa* and so on. Thus generally in the case of such technical terms I avoided translations and used the original Sanskrit words as they are. Of course I have used synonyms whenver they did not create difficulties for me.

Acknowledgement

The present work is a revised version of my Ph.D. thesis. I am grateful to the University of Poona for allowing me to publish it.

I am indebted to Prof. S.S. Barlingay who was my supervisor and guide for my Ph.D. work. He encouraged me in my studies, discussed various philosophical issues with me and made me reconsider my treatment of some of the logical issues tackled in the thesis. Most importantly it was his way of philosophizing which functioned like a guiding star for me. His book 'A Modern Introduction to Indian Logic' had influenced me when I was learning Indian logic as my special paper at M.A. The book inspired me to think of Indian logic in a modern way.

I must gratefully express my gratitude to Prof. M.P. Marathe who initiated me into modern logic. I got the opportunity to attend his lectures on Indian logic and to discuss some issues with him. This provided me with some valuable insights into the problems in Indian logic.

I read with profit some ancient treatises on Indian logic with Pandits like Pt. Kavishwar, Pt. V.B. Bhagwat, Late Pt. Sitaramshastri Kurundkar, Pt. Shrinivasashastri and Pt. Sudhakarshastri Dixit.

My grateful thanks are due to them. I owe a lot to Mr. Avadhut Hardikar who, when he was a high school teacher and I was one of his pupils, initiated my interests in the study of Sanskrit. My grateful thanks are also due to my parents, my wife and children who tolerated and encouraged my involvement in research at various stages of its development.

Amongst the friends of mine the name of Dr. Shrikant Bahulkar should be mentioned who, when he was in Japan, sent to me, a xerox-copy of *Nyāyamukha* translated by Tucci. Without the book my chapter on Diṅnāga would have been thoroughly incomplete.

My grateful thanks are due to Dr. Harishankar Prasad who thought positively about the idea of getting this work published and also to Mr. Sunil Gupta who materialised it in an elegant way.

Poona — Pradeep P. Gokhale

May 1991

Abbreviations

[Figures into brackets usually indicate serial Nos. in Bibliography.]

AMPP *Āptamīmāṁsā Pramāṇaparīkṣā ca* (49)

BHAM *Bhāmatī* (58)

BL *Buddhist Logic* (52)

BSSB *Brahmasūtrabhāṣyam* (50)

BTR *Bhāratīya Tarkaśāstrakī Rūparekhā* (12)

BU *Bṛhadāraṇyakopaniṣad* (1)

CLF *The Concept of Logical Fallacies* (8)

CE Constitutive Element (See p. 38)

CS *Carakasaṁhitā* (2)

CS (1922) *Carakasaṁhitā* (3)

CS (1941) *Carakasaṁhitā* (4)

EIP *Encyclopedia of Indian Philosophies* (45)

FAL *Fallacies* (30)

GK *Gauḍapādakārikā* (27)

GL *General Logic* (24)

HBT *Hetubinduṭīkā* (6)

HIL *A History of Indian Logic* (61)

HL *Hindu Logic as Preserved in China and Japan* (53)

ILA *Indian Logic and Atomism* (35)

ILES *Indian Logic in the Early Schools* (47)

ILSM *An Introduction to Logic and Scientific Method* (17)

JMM *Jaina Darśana Manana Aura Mīmāṁsā* (43)

JSV *Jaina Darśana Svarūpa Aura Viśleṣaṇa* (18)

LT *Laghīyastrayādisaṅgraha* (5)

MIIL *A Modern Introduction to Indian Logic* (10)

MS *Madhyamakaśāstra* (42)

NB *Nyāyabindu* as included in *NBT*

NBh *Nyāyabhāṣya* of Vātsyāyana as included in *NS*

NBT *Nyāyabinduṭīkā* (21)

NBu *Nyāyabhūṣaṇam* (15)

NM *Nyāyamukha* (22)

NMr *Nyāyamañjarī* (31)

NNTI *The Navya-Nyāya Theory of Inference* (41)

NP *The Nyāyapraveśa* (23)

NPS *Nyāyapraveśakasūtram* as included in *NP*

NS *Nyāyasūtrāṇi* (28)

NTK *The Nyāya Theory of Knowledge* (16)

NV *Nyāyavārtika* (55)

NVT *Nyāyavārtikatātparyaṭīkā* (59)

PB *Praśastapādabhāṣyam* (46)

PBTL *Pre-Diṅnāga Buddhist Texts on Logic from Chinese Sources* (54)

PNTA *Pramāṇanayatattvālokālaṅkāra* (60)

PP *Philosophical Papers* (40)

PV *Pramāṇavārtika* (19)

SDS *Sarvadarśanasaṅgraha* (37)

SLRI *A System of Logic Ratiocinatiye and Inductive* (39)

SN *Sāmānyanirukti* (26)

STS *Sammatitattvasopānam* (51)

TAS *Tattvārthasūtra* (57)

TP *A Treatise on Probability* (36)

TS *Tarka-Saṁgraha of Annambhaṭṭa* (7)

TUS *Tattvopaplavasiṁha* (32)

VK *Vaiśeṣikasūtra of Kaṇāda* (33)

VMB *Vyākaraṇamahābhāṣyam* (44)

VN *Vādanyāyaprakaraṇa* (20)

VP *The Vaiśeṣika Philosophy* (56)

VPB *Vākyapadīya (Brahmakāṇḍa)* (14)

VS *Vaiśeṣikasūtra* as included in *VK*

Introduction

§ 0.1

It is human nature to think, to infer for oneself, i.e. to try to know the facts through indirect means when it is not possible or easy to know them through direct means. Man also attempts to convince others about his own knowledge of truths, to justify his knowledge-claims. In doing so he argues. Man also tries to regulate his own way of thinking, his own way of justifying. In doing so he builds the systems of logic in which proper modes of thinking and arguing are studied and distinguished from improper modes.

§ 0.2

In any argument we have two parts. There is a factual claim which is sought for justification. It is called conclusion. The 'grounds' for the conclusion are presented in the form of another part which is called premises. If the premises are judged to be true and they provide logically sufficient ground for the conclusion (i.e. if they entail the conclusion) then the argument is called 'good' or 'sound' argument. But either if any of the premises is judged to be untrue or if the premises do not provide sufficient ground for the conclusion, or both, then the argument is called bad or unsound argument. When premises, whether true or false, provide logically sufficient ground for the conclusion, then the argument is called valid, but it is not called sound unless also the premises are judged to be true. Similarly irrespective of whether premises are true or false, if the premises do not provide sufficient ground for the conclusion, then the argument is called invalid. Every invalid argument is unsound but not vice versa.

§ 0.3

But some modes of thinking and justifying are such that they are not supposed to be characterised either as valid or invalid. The reasoning that is involved in inductive generalisation or the one that is involved in analogical argument is not formally valid, because the premises in such a reasoning do not entail the conclusion; they do not provide sufficient ground for the conclusion. But when such a reasoning in made with the clear idea about the limitations of such a type of reasoning, the questions are not raised as to the validity or invalidity of that reasoning, but rather as to the strength and weakness of that reasoning. And if such a reasoning has a defect which weakens the reasoning then that defect is called a fallacy; that reasoning is called fallacious analogical reasoning or fallacious inductive reasoning. Here it will be useful to distinguish between valid, sound and proper. We have already noticed the distinction between valid and sound. Regarding a given analogical reasoning (when there is a clear understanding that it is analogical reasoning) the question as to its validity or soundness is not raised, but the question whether the reasoning is proper or improper can be raised. The question pertaining to propriety of analogical reasoning is really the question pertaining to strength of that analogical reasoning. Here we are using the words 'proper' and 'improper' more vaguely than the words sound and unsound. Every sound argument is a proper argument. But every proper argument (for instance proper analogical reasoning) is not a sound argument. Instead of proper and improper one may choose the words good and bad or correct and incorrect with the same vagueness and distinguish the notions of these from those of sound and unsound.

The proper and improper modes of thinking and arguing that were discussed in Indian logic upto Diṅnāga and Udyotakara were the modes of analogical reasoning (although there was not very clear understanding that they were the modes of analogical reasoning only). Dharmakīrti was the first Indian logician to consider seriously the sound and unsound, valid and invalid modes of reasoning. Naiyāyikas who succeeded Dharmakīrti mixed between these two considerations: consideration about analogical reasoning, which was more practical than logical and consideration about validity and soundness of reasoning which was more logical than practical. In this way the notion of *anumāna* as *pramāṇa* underwent a change and this had an impact on the history of the concept of fallacy also.

§ 0.4

Although various philosophical systems flourised in India after the Upanishadic period and almost all of them dealt with the problems of knowledge and justification, the main amongst those who tried to develop the systems of logic were Nyāya and Buddhism. In fact one of the remarkable features of the history of Indian logic is the conflict between Nyāya and Buddhist theories of logic. Indian logic partly developed through this conflict and partly remained static because of the revivalist and conservative tendencies amongst the logicians. One of our aims in this treatise is to draw a sketch of this development as well as resistence to development in Indian logic with special reference to the concepts of inference and fallacy. I believe that such a historical approach to Indian logic as against the traditionalistic approach proves fruitful in some important ways. For instance traditionally it is held that there is no basic change in the Nyāya concepts of inference and fallacy from Akṣapāda to Vācaspatimiśra and even to the Navya-Nyāya logicians. Some modern scholars also, under the influence of this traditional approach, have held that the change was only in the emphasis. Early Naiyāyikas emphasised the aspect of material truth, whereas the later Naiyāyikas emphasised that of formal validity and conceptual consistency.[1] Similarly it is generally held that there is no basic difference between Diṅnāga's and Dharmakīrti's notions of inference and fallacy. Traditionally we study the systems as homogeneous static wholes and thus we disregard any differences between the views of the philosophers which belong to the 'same' system. On the other hand if we examine the systems historically, we will be able to know how the systems interacted with each other and underwent a change in the course of history. By doing such an exercise with reference to the concepts of inference and fallacy we can see how the concept of inference as *pramāṇa* and the corresponding concept of fallacy underwent a change even within the Nyāya tradition. We also notice, for instance, that Diṅnāga's concept of inference was basically different from Dharmakīrti's concept.

§ 0.5

Under the direct or indirect influence of the western conception of fallacies, some orientalists have equated the notion of '*hetvābhāsa*' with the notion of 'logical fallacy' (i.e. formal fallacy). BODAS, for instance, in his notes to *Tarkasaṅgraha* has tried to compare Aristotelian fallacies of syllogism with *hetvābhāsas.*[2] NANDITA BANDYOPADHYAY too seems to have done so with reference to the Navya-Nyāya concept of *hetvābhāsa.*[3] These attempts are misled because the notion of *hetvābhāsas* as discussed in Navya-Nyāya has some logical links with the notion of material fallacies discussed in western logic; but at any rate *hetvābhāsas* are not formal fallacies. When one identifies some *hetvābhāsa* in a given inference, one is pointing out the falsehood or dubitability of some premise or of the conclusion. We are not pointing out that the inference is invalid. Here there is a slight difference between the ancient Indian concept of *hetvābhāsa* and the Navya-Nyāya concept.[4] In the former *hetvābhāsa* is defined as the defective *hetu* or rather the pseudo-*hetu*, whereas in the latter it is defined as a defect *in the pseudo-hetu*. In both the cases it is held that falsehood or dubitability of some premise or that of the conclusion ultimately makes the *hetu* defective. Hence both the notions of *hetvābhāsa* are logically linked with the notion of 'material fallacy'.

Such attempts as those of BODAS and BANDYOPADHYAY also represent the tendency to regard Navya-Nyāya as *the* representive of Indian logic. This tendency is misleading not only because the ancient concept of *hetvābhāsa* is different from Navya-Nyāya concepts of *hetvābhāsa* but also because the study of fallacies in the light of Navya-Nyāya account would give us only an incomplete and misleading picture of Indian account of fallacies. When we study ancient Indian logic, we find that it had a broader notion of fallacies as compared with that of Navya-Nyāya logic. Many varieties of fallacy which the ancient Indian logicians discussed were rather neglected by Navya-Nyāya logicians. I propose in this work to concentrate upon fallacies discussed in ancient Indian logic of pre-Udayana period. Although separate consideration of Navya-Nyāya concept of fallacy and the consideration of its relationship with the ancient conception would be an important topic of research, it is beyond the scope of this treatise.

§ 0.6

Those who are aware of this fact that the notion of *hetvābhāsa* is basically a notion of material fallacy, sometimes may go to the other extreme and suggest that Indian logicans did not develop any notion of formal fallacy; rather they did not discuss the question of formal validity and invalidity at all. But as we shall see, Dharmakīrti did consider the problem of validity seriously and dealt with some formal fallacies under the head *'dṛṣṭāntābhāsa'*. However, it has to be noticed that validity of an inference cannot be identified with *'prāmāṇya'* of that inference. When a particular inference was held to be *pramāṇa* in Dharmakīrti's logic it was also held to be valid, but validity was only a necessary condition of *parārthānumāna's* being *pramāṇa* and not the sufficient condition. The sufficient condition was soundness which involved both validity and truth of the premises. Arriving at or showing to have arrived at true conclusion through a valid argument with true premises was the goal in using *anumāna pramāṇa* (especially *parārthānumāna*). So the consideration of validity never assumed an isolated place in the discussions of *anumāna* in Indian logic.

Notes

1. This view has been held, for instance, in *NTK* (See its Introduction, p. 5).
2. See, 8.17.
3. BANDYOPADHYAY entitles her book 'The Concept of Logical Fallacies' but devotes it to the study of Navya- Nyāya concept only. She also says, 'The *hetvābhāsas* or the fallacies of the middle term are the logical fallacies in the proper sense of the term ....' (Intro. to *CLF*, p.i.).
4. Ancient Indian concept of *hetvābhāsas* was that of 'a *hetu* which seems to be a sound one but is not' (See Note 2 under Chapter VIII). But the Navya-Nyāya notion of fallacies as

BANDYOPADHYAY points out is that of an object of a true cognition, which contradicts with some cognition *(parāmarśa* or *anumiti)* involved in the inferencial process. In other words, according to Navya- Nyāya, fallacy (rather, ***hetvābhāsa***) is a fact which falsifies some premise or the conclusion of the given inference. This idea is certainly different from the ancient idea of *hetvābhāsa*. BANDYOPADHYAY (badly) describes this 'falsifying fact' as 'the object of a valid contradictory knowledge'. See for instance, Intro. to *CLF*, p. ii.

Chapter I

The Nature of Inference and Fallacies Accepted by Pre-Akṣapāda Indian Logicians (Especially by Caraka)

§ 1.1

The history of committing fallacies is as old as the history of human reasoning. The awareness of the demarcation between sound and unsound arguments came later.

The dialogue between Gārgī and Yājñavalkya described in *Bṛhadāraṇyaka*[1] is worth mentioning in this context. Gārgī asks Yājñavalkya, "What is that in which the Water is woven in, as all this world is woven in the Water?" Yājñavalkya answers, "It is woven in the Air". Again Gārgī asks, "What is that in which the Air is woven in?" Yājñavalkya answers, "It is woven in the worlds of Sky." But Gārgī is not satisfied. She goes on asking the same type of question, "What is that in which A is woven in?" And Yājñavalkya answers in the similar way, "It is woven in B". The danger is that this chain of answers (A is woven in B, B is woven in C etc. etc.) may lead to infinity. By asking the same type of question Gārgī was as if pointing out that such a chain of metaphysical enquiry will probably lead us to the fallacy of infinite regress. And when Yājñavalkya at last warned Gārgī by saying that she should not ask more, otherwise her head would fall down; he was as if suggesting that if we want to avoid infinite regress, we have to stop somewhere. No term synonymous to 'infinite regress' occurs in this Upaniṣadic dialogue. But this is how the awareness about the faults of thought was appearing in the minds of Indian thinkers.

The Upaniṣadic philosophizing, however, developed through analogies rather than inferences and it took some centuries to have distinguished between *anumāna* and *jāti* (an argument based on only similarity or difference). The awareness of fallacies is seen to be developed a little at the time of *Mahābhārata* in which Nārada is described as the knower of the virtues and the faults of the five-membered statement.

(पञ्चावयवयुक्तस्य वाक्यस्य गुणदोषवित्[2])

§ 1.2

Caraka is perhaps the first thinker, whose discussion on the nature and possible faults of controversy is available. In *CS, Vimānasthāna, Adhyāya* 8, Caraka gives some useful instructions to the physicians regarding the method of controversy. He mentions forty four terms that should be understood for knowing the manner of controversy (वादमार्गज्ञानार्थम् अधिगम्यानि पदानि). The list includes the metaphysical principles like *dravya, guṇa, karma* etc., the means of knowledge, the members of inferencial statement, the fallacies of *hetu (ahetu)*, the faults of statement (*vākyadoṣa*), the occasions of defeat (*nigrahasthāna*), the refutations based on similarity or difference (*uttara*) and the refutations based on ambiguity of language (*chala*) as well. While giving the theory of debate Caraka distinguishes between two kinds of debate namely *Sandhāya saṁbhāṣā* and *Vigṛhya sambhāṣā* (which may be translated respectively as friendly debate and hostile debate). One of the main differences between the two is that in friendly debate, the fallacies are discovered but tolerated,[3] but in the hostile one, discovering a fallacy committed by the other debater just amounts to defeating him.

§ 1.3

Caraka gives several instructions to the debaters who want to get success in the hostile debate. The ways of success suggested by Caraka include the prior knowledge of the proficiencies and drawbacks of the opponent, attacking the opponent's weak points in the debate, exhibiting the strong points of oneself which cannot be easily overthrown, and many other tricks. The hostile controversy is described by Caraka like a battle which is never played liberally.

Even the non-intellectual and non-academic ways like using difficult constructions of language, smiling in between like a victor, speaking continuously without giving any chance to the other debater, blaming his standpoint without giving reasons, blaming the intellectual standard and scholarship of the opponent etc. are prescribed by Caraka if a foolish opponent is to be tackled. He also suggests the various ways of taking advantage of the opponent's intellectual and mental weaknesses.[4]

Here the distinction between academic and non-academic ways of winning a controversy is important. The non-academic ways can again be classified into two classes. One is the class of those tricks which are used to harass the opponent than to refute his thesis. Such a way of defeating can be called as non-academic as well as non-intellectual. According to this way, the disputant does not attack the argument of the opponent but he attacks the opponent. The other class is the class of faulty statements and unsound arguments used for creating the impression that the opponent's argument is fallacious. *Chala*, which Caraka mentions in his above-mentioned list of terms, is a case of non-academic but intellectual way of achieving success in the debate. The ways like using difficult constructions of language, blaming the other party without giving reasons etc. are non-intellectual in nature. They have psychological bearing and not the intellectual one. Such non-intellectual ways of defeat which are mentioned and supported by Caraka do not find mention in the later treatises on the method of debate.

The roots of Akṣapāda's notion of *jāti* (which means a refutation by arguing on the basis of mere similarity or difference) are found in Caraka's notion of *uttara*. By *uttara* Caraka means 'stating the difference (between the positive instance and the instanciated) when the debater is arguing on the basis of similarity (between them) or stating the similarity (between the negative instance and the instanciated) when the debater is arguing on the basic of difference (between them)'.[5]

Nigrahasthāna is defined by Caraka as 'getting defeated'.[6] *Nigrahasthāna* literally means place or occasion of defeat. Caraka made a list of fifteen occasions of defeat. These fifteen occasions of defeat are varied in nature. Some of them are fallacies, some are linguistic mistakes, some are just the breaches of the conventions of controversy. Caraka enlists them without classification.[7] Akṣapāda raised the list by seven by either adding to or splitting the members of the Caraka's list. KASHINATH SHASTRI and GORAKHANATH CHATURVEDI point out that only four occasions of defeat namely *aprāptakāla, ananubhāṣaṇa, apratibhā* and *vikṣepa* are the additions of Nyāya.[8] But in fact *aprāptakāla* mentioned by Akṣapāda is a restricted form of *kālātītavacana* mentioned by Caraka. *Kālātītavacana* refers to any mistimed statement of the debater, while *aprāptakāla* refers to disordered and, in this sense, mistimed presentation of the members of the inferencial statement. *Ananubhāṣaṇa* and *apratibhā* were also suggested by Caraka when he said that *nigrahasthāna* is 'not understanding the statement uttered by the other debater thrice'.[9]

§ 1.5

Caraka defines *anumāna* as *'tarkaḥ yuktyapekṣaḥ'*[10] which means the speculation based on the application of *hetu*. Thus *anumāna* for him is more a speculation than an inference proper. His examples of *anumāna* are stated by him as, 'Abdominal fire is inferred from the digestive power of an animal, the strength is inferred from the capacity to work, the existence of sense-organs is inferred from the cognitions of sense-data'.[11] These examples express the judgements regarding the imperceptible objects. Analogy plays unavoidable role in judgements about such objects. Such *anumānas* were later on termed as *sāmānyato dṛṣṭa anumāna*.

Carakasaṁhitā is the first available source in which the five-membered inferencial statement is exemplified.[12] The first member was the statement of thesis (*pratijñā*) and the proof (*sthāpanā*) of that thesis consisted of the remaining four members namely, *hetu, dṛṣṭānta, upanaya* and *nigamana*.[13]

Caraka's example of five-membered inferencial statement is as follows,

Pratijñā: Puruṣa is permanent.

Hetu: because of being non-created.

Dṛṣṭānta: like space.

Upanaya: Just as space is non-created and it is permanent, so is *Puruṣa*.

Nigamana: Therefore (*Puruṣa*) is permanent.[14]

We find that in this example space and *Puruṣa* (meaning soul) are the individuals which are analogous in character. The signifying analogy is 'being non-created' while the signified analogy is 'being permanent'. In *upanaya*, the instance, namely the space, is said to be characterised by both the analogies. In *hetu, Puruṣa* is said to possess the signifying analogy. But the universal statement of the form 'whatever is non-created is permanent' is absent.

In some books of *Carakasaṁhitā*, there is alternative reading of *dṛṣṭānta* and *upanaya* quoted. It is as follows,

Dṛṣṭānta:Space is non-created and also permanent.

Upanaya:As space is non-created. So is *Puruṣa*.[16]

This reading seems to be more correct than the first. The first reading of *upanaya* gives an impression that both signifying analogy and signified analogy are known to co-exist in the *sādhya*-case; that is, *Puruṣa* is known to be both non-created and permanent. And this is certainly false. Anyhow, any of the two readings do not give any indication of the inclusion of *vyāpti*-statement as a premise.

Caraka being primarily a scientist, was very much concerned with the general statements expressing causal connection between things or events. The insight in the causal connections was, according to him, an important tool of examining any thesis. In *Sūtrasthāna*, Caraka calls this insight as *yukti*.[17] It is also possible that in Caraka's definition of *anumāna*, namely '*tarkaḥ yuktyapekṣaḥ*', *yukti* means the insight stated above. However, Caraka does not clarify this point, nor does he emphasise it. The inferencial statement exemplified by Caraka, as we have already seen, is clearly a statement of analogical reasoning. Thus Caraka's account of *anumāna* seems to be ambivalent between the scientific account regarding the universal causal statements and the traditional account of *anumāna* as analogical reasoning. The scientific insight shown by Caraka was not, however, pursued by the successive *Nyāya* logicians.

§ 1.6

Under the class of fallacies of *hetu (ahetu)* Caraka includes *prakaraṇasama, saṁśayasama* and *varṇyasama*, out of which only *prakaraṇasama* was mentioned by name by *Akṣapāda* in his list of *hetvābhāsas*. The mention of *saṁśayasama* and *varṇyasama*, is found in Akṣapāda's list of *jātis*. But *saṁśayasama* of *Akṣapada* is very different from Caraka's *saṁśayasama*. Let us discuss Caraka's fallacies of *hetu* one by one.

Prakaraṇasama : While explaining the example of *prakaraṇasama*, Caraka says, " ..... this is not a sound *hetu*. Because the *hetu* cannot be the same as the proposition to be proved."[18] Thus Caraka's *prakaraṇasama* is clearly a form of begging the question. The example of the fallacy given by Caraka is as follows,

"As the soul is (altogether) different from body, it is eternal; because body is not eternal and soul must have different characteristics".[19]

Here the arguer has assumed the truth of the reason that the soul is distinct from body in all its respects. But the reason itself needs the proof. If it is proved that the soul is distinct from body in all its respects, it follows that soul must be eternal. But the proof of the reason, namely the soul is distinct from body in all its respects, depends on the proof of the conclusion, namely the soul is eternal.

Saṁśayasama : Caraka defines this pseudo- *hetu* as one which is in fact doubtful (as to prove the *sādhya*) but is asserted for removing the doubt. The example runs as follows :

"When there is a doubt about some person as to whether in spite of his partial knowledge of medical science, he is a medical practitioner or not; the arguer argues that because he knows a part of medical science, therefore he is a medical practitioner."[20]

The example suggests that *saṁśayasama* also is a form of question-begging, because here the question that has to be answered is assumed to have answered itself. The difference between *prakaraṇasama* and *saṁśayasama*, however, is that in case of *saṁśayasama* the doubt that is assumed to be removed is about the relation between *hetu* and *sādhya*, when in case of *prakaraṇasama* the doubt is about the existence of *hetu* in *pakṣa*-case.

Saṁśayasama can be identified with a form of *savyabhicāra* of the later Nyāya tradition where the concomitance (in the observed world) between *hetu* and *sādhya* is doubtful. But the better interpretation of both, *prakaraṇasama* and *saṁśayasama* is that in both the cases the method of analogy is not applied at all, when it should have been applied. Here the case is argued without giving any ontic evidence (that is, an analogous case where the signifying analogy and signified analogy are found co- existent or co-absent), and when the arguer is asked for evidence, the *hetu* or the relation between *hetu* and *sādhya* is again asserted and thus the question is begged. This interpretation seems to be more correct especially in the context of Caraka's form of inferencial statement as a statement of analogical reasoning. It may be noted that Caraka's form of inferencial statement is the weakest form of analogical reasoning as compared with the forms discussed by Akṣapāda and the later logicians.

Varṇyasama : In case of *varṇyasama*, the instance (*dṛṣṭānta*) given by the arguer is unable to support his argument. Thus it is a fault in the analogous case than in the *hetu*. The example is as follows,

"the intellect (*buddhi*) is impermanent because it cannot be touched, like the sound."[21]

Here the instance namely sound is doubtful as to be impermanent. Thus it cannot support the argument. Caraka says, "Here the sound is subject to explanation (*varṇya*) and the intellect also is subject to explanation. Thus the *hetu* being commonly applicable for the proof of both the disputable cases (namely *dṛṣṭānta*-case and *sādhya*-case) is not a (good) *hetu*."[22]

Akṣapāda mentions *varṇyasama* in the list of *jātis*. When an opponent commits *varṇyasama jāti*, he (wrongly) argues that the instance given by disputant is itself disputable just as the thesis is disputable.[23] Thus *varṇyasama jāti* can be called as 'a (wrong) claim of Caraka's *varṇyasama*'.

Varṇyasama can be included in the list of *dṛṣṭāntābhāsas*. The above example of *varṇyasama*, according to Dharmakīrti's list, is an example of *sandigdhasādhyadharma*.[24]

§ 1.7

The term *savyabhicāra* has been mentioned by Caraka as one of the terms relevant in debate. But *savyabhicāra* in *Carakasaṁhitā* does not refer to a kind of unsound *hetu*, but to anything which is related with something else irregularly or with uncertainty. His example is, 'This medicine may be useful in this decease or not'.[23] Here the relation between the particular medicine and the cure of the particular decease is irregular or uncertain. He defines *savyabhicāra* as *yadvyabhicaraṇam*,[26] which means, 'that thing or property which deviates from the other thing or property'. This general notion of *savyabhicāra* seems to have been employed by Akṣapāda in his classification of *hetvābhāsas*.

Caraka includes *viruddha* under the faults of statement (*vākya-doṣa*). Any statement made by the debater which is contrary either to the example he is presenting to support his thesis, or to some accepted doctrine of the debater or to the general presuppositions of the topic is regarded as the *viruddha* statement by Caraka. The above-mentioned varieties of *viruddha* are called *dṛṣṭāntviruddha*, *siddhāntaviruddha* and *samayaviruddha* respectively. Akṣapāda's *viruddha- hetvābhāsa* could be regarded as a fallacy of *hetu*-statement and in that case it would resemble *siddhānta-viruddha* of Caraka. It is possible, however, that *viruddha-hetvābhāsa* was Akṣapāda's own contribution.

Kālātīta-vacana or *atītakāla* of Caraka can hardly be assimilated with *kālātīta-hetvābhāsa* of Akṣapāda. *Kālātīta-hetvābhāsa* must not be concerned with the mistimed presentation of inferencial statements, but it must be concerned with the confusion about the temporal distances between the occurence and appearance of *hetu* and *sādhya*.[27] *Kālātīta-hetvābhāsa* does not necessarily presuppose the context of debate while *kālātīta- vacana* of Caraka does. Caraka's *kālātīta-vacana* or *atītakāla* resembles Akṣapāda's *aprāptakāla*.[28]

§ 1.8

VIDYĀBHŪṢAṆA points out that prior to Caraka, Medhātithi Gautama, a logician referred to in *Mahābhārata* and Bhāsa's *Pratimānāṭaka*, had laid out certain rules of controversy and logic, which Caraka borrows in his *Carakasamhitā*. This is certainly an important observation. It is, however, sufficient for our purpose to note that Caraka's account of the method of debate, which appeared some three centuries before Akṣapāda, includes some thoughts on inference and fallacies. As the Nyāya-school developed, we find that, the Akṣapāda's account was wrongly interpreted in the light of the logical theory of later Naiyāyikas. But Akṣapāda's account of inference and fallacies resembles more in form and content with the earlier account of Caraka than with the developed account of the later logicians like Jayanta and Vācaspatimiśra. That is why, for understanding Akṣapāda's account, it is very useful to study Caraka's account.

Notes

1. *BU* 3.6.
2. Quoted by VIDYĀBHŪṢAṆA, *HIL*, p. 43 (footnote 3).
3. न च निग्रहभयाद् उद्विजेत् निगृह्य चैनं न हृष्येत् । *CS*, p. 744.
4 श्रुतहीनं महता सूत्रपाठेन अभिभवेत्, विज्ञानहीनं पुनः कष्टशब्देन ...... एवमेतैरुपायैः परमवरमभिभवेत् शीघ्रम् । *CS*, p. 747.

The similar suggestions can be found in the following passage from Aristotle's Sophistical Refutations,

"With a view then to refuation, one resource is length - for it is difficult to keep several things in view at once; ..... One resource, on the other hand, is speed; for when people are left behind they look ahead less. Moreover, there is anger and contentiousness, for when agitated everybody is less able to take care of himself.

Elementary rules for producing anger are to make a show of the wish to play foul, and to be altogether shameless." (Quoted in *FALLACIES*, pp. 51-2).

5. See *CS*, pp. 751-2.
6. *'parājayaprāptiḥ'*, *Ibid*, p. 762.
7. In fact there can be doubt regarding the propriety of some of the 'occasions of defeat'. We shall discuss this problem in § 7.63 - 70.
8. See *CS*, p. 763.
9. *Ibid*, p. 762.
10. *Ibid*, p. 753.
11. *Ibid*, p. 753.
12. Bhadrabāhu, a Jaina philosopher, who flourished probably before Caraka, is seen to have described a ten-membered inferencial statement (See *HIL*, pp. 166-167). In the ten members, *upanaya* from the popular five members is omitted and other six members namely *pratijñāvibhakti, hetuvibhakti vipakṣa, vipakṣapratiṣedha, āśaṅkā* and *āśaṅkāpratiṣedha* are added. *Pratijñāvibhakti* and *hetuvibhakti* serve as the clarified forms of *pratijñā* and *hetu* respectively. *Vipakṣa* is an objection in which a concrete case of (possible) *vyabhicāra* is shown. In *vipakṣapratiṣedha* the possibility of the particular *vyabhicāra*-case stated in *vipakṣa* is ruled out. *Āśaṅkā* serves as the objection against the propriety of the *dṛṣṭānta*-case stated in *dṛṣṭānta*, and *āśaṅkāpratiṣedha* serves as the removal of the objection. In all the ten members, however, the statement of *vyāpti* is absent.
13. अथ प्रतिज्ञा नाम साध्यवचनं यथा नित्यः पुरुष इति । अथ स्थापना । स्थापना नाम तस्या एव प्रतिज्ञायाः हेतुदृष्टान्तोपनयनिगमनैः स्थापना । *CS*, p. 750.
14. यथा नित्यः पुरुष इति प्रतिज्ञा । हेतुः अकृतकत्वादिति । दृष्टान्तः यथाकाशमिति । उपनयः यथा चाकृतकमाकाशं तच्च नित्यं तथा पुरुष इति । निगमनं तस्मान्नित्य इति । *CS*, p. 750.
15. I have borrowed the expressions 'signifying analogy' and 'signified analogy' from COHEN and NAGEL. See *ILSM*, p. 287. Also see § 2.3.
16. दृष्टान्तो ऽकृतकमाकाशं तच्च नित्यम्, उपनयो यथा चाकृतकमाकाशं तथा पुरुषः । *CS* (1941), Footnote No. 2 on the first column of p. 267.
17. बुद्धिः पश्यति या भावान् बहुकारणयोगजान्, युक्तिस्त्रिकाला सा ज्ञेया .... । *CS*, p. 221.
18. सोऽयमहेतुः । न हि य एव पक्षः स एव हेतुः इति । *CS*, p. 756.

    The word *'pakṣa'* here, is used in the sense of the proposition to be proved. In ancient Indian logic *pakṣa* generally means the view of the debater which is being established or his proposition that is to be proved. Thus *pakṣa* had epistemological or linguistic status.

Gradually in the later Nyāya logic, however, *pakṣa* was given an ontological status. MATILAL seems to have observed this. See his article 'A Note on Nyāya Fallacy Sādhyasama and Petitio principii' published in JIP, Vol. 2, Nos. 3-4, p. 215.

19. See *CS*, p. 756.
20. *Ibid.*
21. *Ibid.*
22. *Ibid.*
23. See § 7.11.
24. In case of this fallacy, it is doubtful whether the *sādhya*-predicate can be predicated of the *dṛṣṭānta*-subject or not. See § 5.27.
25. See *CS*, p. 754.
26. *Ibid.*
27. For *kālātīta* or Akṣapāda, see § 2.21-23.
28. See § 1.4.

Chapter II

The Nature of Inference and Hetvābhāsa described by Akṣapāda and Vātsyāyana

Part I

§ 2.1

In ancient Indian logic, the term *pramāṇa*[1] was being used generally in the sense of authentic knowledge. The term *pratyakṣa* also was used mainly in the sense of true and doubtless perception and not in the sense of the means of such perception. The same was the case with *upamāna*. Similarly *anumāna* also meant the true cognition which was preceded by perception and did not generally mean the means of such cognition. What should be called the means of such cognition, whether *hetu* or *hetujñāna* or *vyāpti* or *vyāptijñāna*, was not the problem before the ancient logicians. The problem before them seems to be, how to distinguish between *mithyājñāna* (false cognition or doubt) and *pramāṇa* (true and doubtless cognition).

The two terms in Akṣapāda's definition of *pratyakṣa*,[2] namely *avyabhicāri* and *vyavasāyātmaka* apply not only to *pratyakṣa* but to all *pramāṇas*. The term *vyavasāyātmaka* means 'assertive' or 'involving no doubt', while *avyabhicāri* means 'not-false'. If the cognition of the form p is to be taken as *pramāṇa*, then modern logicians may argue that 'p-or-q' is also a form of true cognition. But Indian logicians would not call it a form of *pramāṇa* because p-or-q is not *vyavasāyātmaka*.[3] The cognition of the form not-p, in that case, would be *vyavasāyātmaka*, but false, and hence *apramāṇa*, because not-p amounts to *vyabhicāra* or deviation from the true cognition which is of the form p.

Thus the term *vyavasāyātmaka* is concerned with the form of the cognition while the term *avyabhicāri* is concerned with the content of it. The condition that *pramāṇa* should be *vyavasāyātmaka* means that the form of the cognition should not be disjuctive. The condition that *pramāṇa* should be *avyabhicāri* means that the content of such cognition should obtain as a matter of fact. That is to say, the content should not deviate from the real state of affairs. Thus to say that *anumāna* should be *avyabhicāri* is to say that the object of *anumāna* should obtain as a matter of fact. If we interpret *vyabhicāra* in this general manner, *vyabhicāra* could cover almost all the *hetvābhāsas*.[4]

From the above discussion, we can see that when *anumāna* was regarded as *pramāṇa* it was assumed that we can determine the conditions in which the particular *anumāna* can give true results, i.e., it can be *avyabhicāri*. However, whether Akṣapāda and Vātsyāyana could really determine such necessary and sufficient conditions of a reliable inference, is a different problem. We shall discuss in the following lines whether Akṣapāda's and Vātsyāyana's frame of *anumāna* was only that of analogical reasoning (which cannot be completely reliable) or that of the sound reasoning (which is valid and has true premises, and thus is reliable).

§ 2.2

Before we come to the actual examination of Akṣapāda's and Vātsyāyana's account of *anumāna*, it would be useful to clarify the major aspects of analogical reasoning.

Analogical reasoning is made from particular(s) to particular(s). The particular or particulars from which the analogical reasoning is made can be called as the instance (s) or *dṛṣṭānta(s)* of the analogical reasoning. The instances may be either positive or negative in which case they are called *sādharmyadṛṣṭānta* and *vaidharmyadṛṣṭānta* respectively. The particular or the particulars about which the reasoning concludes, are called as the *sādhya*-case. The *sādhya*-case may be singular or plural. The similarity or difference, i.e., the common or distinguishing property, which exists in *sādhya*-case, and on the basis of which the conclusion is drawn is called signifying analogy. The similarity or difference, i.e. the common or distinguishing property which is being proved to exist in *sādhya*-case is called signified analogy.[5]

This signifying or signified analogy may be either positive analogy or negative analogy.[6] That is to say, the argument from analogy may be based either on observed constant similarity between the *sādhya*-case and the positive instances, or on observed constant difference between the *sādhya*-case and the negative instances. Sometimes the argument may be based on both.

When some unknown similarity between instances and *sādhya*-case is proved on the basis of some known similarity between them, the similarity or the common property that is being proved to exist in *sādhya*- case, may be called positive signified analogy. The common property on the basis of which the positive signified analogy is proved to exist in *sādhya*-case may be called positive signifying analogy.

Similarly when some unknown difference between instances and *sādhya*-case is proved on the basis of some known difference between them, the distinguishing property which is known not to exist in the negative instances, but is not known to exist or not to exist in *sādhya*-case, and which is being proved to exist in *sādhya*-case can be called the negative signified analogy. The distinguishing property which is known not to exist in the negative instances, but is known to exist in *sādhya*-case and on the basis of which the negative signified analogy is being proved, is called the negative signifying analogy.

Thus the positive instance is known to have the signifying and the signified analogy together, and the negative instance is known to have the absence of both. On the other hand, the *sādhya*-case is known to have signifying analogy but not known to have the signified analogy. The conclusion of such reasoning conveys that *sādhya*-case also has signified analogy as a property.

The analogical reasoning does not have universal statement as its premise. It has only the existential statments as its premises which assert the following:

1. The cases in which the signifying and the signified analogies co-exist and/or the cases where both the analogies are absent.

2. The existence of the signifying analogy in the *sādhya*-case.

The conclusion asserts the existence of the signified analogy in the *sādhya*-case.

If any case is obtained where there is the signifying analogy but not the signified analogy, the probability of the reliability of the analogical reasoning is reduced.

§ 2.3

In fact any so-called inductive reasoning is a modification of analogical reasoning. For instance, when somebody observes hundred human beings and also observes that they are mortal, then before arriving at the so-called inductive conclusion that all men are mortal, the logically necessary step is that all the remaining individuals which are human beings but whose mortal nature is not known, must be mortal. And this step is the conclusion of analogical reasoning. We get the inductive conclusion just by conjuction of the following statements.

1. The hundred human beings are mortal.

2. The remaining human beings must be mortal.

The first statement, in this context, is an observational statement, while the second statement is the conclusion of the analogical reasoning in which,

1. 'hundred human beings' are positive instances

2. 'the remaining human beings' is the *sādhya*-case

3. 'being human' is the (positive) signifying analogy.

4. 'being mortal' is the (positive) signified analogy.

Thus it may be seen that empirical generalisation is a combination of observation and analogy. Interpreted in this way, the process of induction is not of the form 'particular -- general', but of the form 'particulars (observed) -- particulars (unobserved) -- general'. In this process the first part, namely 'particulars (observed) -- particulars (unobserved)' is analogical reasoning. Similar to this analogical reasoning, every analogical reasoning is a pre- inductive reasoning (or rather pre-'inductive- generalisation' reasoning) and may be distinguished from the deductive reasoning which is post-inductive reasoning. My contention here will be that the ancient Indian logicians before Dharmakīrti conceived of *anumāna* as pre- inductive reasoning while the logicians starting from Dharmakīrti conceived of it as post-inductive reasoning.

In this way, many problems which are raised regarding the logical strength of the method of induction are really the problems regarding the logical strength of the analogical reasoning. Analogical reasoning or empirical generalisation is supposed to hold good as far as a contrary case or *'vyabhicāra'* is not found. Thus *vyabhicāra* provides a negative check for the strength of analogical reasoning as well as for that of empirical generalisation. It should be remembered, however, that *vyabhicāra* does not provide a sufficient check, so that just the non-availability of *vyabhicāra* would prove the conclusion of the analogical reasoning or of the inductive reasoning.

§ 2.4

In *Nyāya-sūtra* of Akṣapāda, *anumāna* is referred to as a doubtless true cognition *(pramāṇa)* and not just a *tarka.*

Akṣapāda clearly mentions that *anumāna* is not only of the *sāmānyato dṛṣṭa* kind (i.e., based on simple similarity and used to judge some imperceptible object[8]) but it may be of the *pūrvavat* or *śeṣavat* kind.[9] Generally the two-fold classification of *anumāna* was made into *dṛṣṭa* and *sāmānyatodṛṣṭa* kinds. Akṣapāda seems to have split the *dṛṣṭa* type of inference into two classes namely *pūrvavat* and *śeṣavat*. Thus Akṣapāda meant that a major class of *anumānas* refers to the *anumānas* which have the empirically verifiable conclusion. Thus the scope of *anumāna* seems to have been seen more clearly by Akṣapāda than by Caraka. Caraka had considered only positive analogies while describing *anumāna* while Akṣapāda added negative analogies as well.

§ 2.5

While defining the members of inferencial statement, Akṣapāda again and again refers to *sādharmya* and *vaidharmya*. Even *hetu* has been defined by Akṣapāda as the means of proving *sādhya* with the help of similarity with the example or difference from the example.[10] Here *hetu* means nothing but 'signifying positive analogy' in the first case and 'signifying negative analogy' in the second. Instance *(dṛṣṭāta)* is defined by Akṣapada as the case analogous to *sādhya*, either positively or negatively.[11] *Upanaya* which literally means 'bringing near (the conclusion)' has been defined by Akṣapāda as 'gathering *(upasaṁhāra)* the sādhya-case together with the instance (by demonstrating the similarity or difference) by *tathā* or *na tathā* (respectively).'[12] The term *'tathā'* in the *upanaya*- statement cannot be taken to mean just the conjunction (of *hetu*-statement and *dṛṣṭāta*-statement) but seems to function as a relational property which means 'is (positively) analogous to'. Similarly *'na tathā'* in the *upanaya*- statement means 'is negatively analogous to'.[13] For a valid presentation of post-inductive inference, it is necessary to assert the universal relation between *hetu* and *sādha*, while Akṣapāda was satisfied with the relation of 'being analogous to' between sādhya-case and instance. Akṣapāda's *anumāna* was therefore a pre-inductive reasoning. As shown in § 2.3, the fallacy called *savyabhicāra* in *Nyāyasutra* functioned as a negative check for the strength of *anumāna*, and the mention of *savyabhicāra*, did not give *anumāna*, the status of post- inductive sound inference.

§ 2.6

Let us now deal in brief with Vātsyāyana's conception of *anumāna*. Vātsyāyana emphasises the relation between *liṅga* and *liṅgin*, i.e. signifying analogy and signified analogy, rather than just the similarity or difference between sādhya-case and *dṛṣṭānta*-case. While analysing the term *'tatpūrvakam'* in Akṣapāda's definition of *anumāna* (*NS*, 1.1.5) he says that it is the observation of the relation between *liṅga* and *liṅgin* before the time of inferring and the observation of *liṅga* in the *sādhya*-case at the time of inferring that is meant by

Akṣapāda by the word *'tat'* and *anumāna* is preceded by these two sorts of observations.

Vātsyāyana many times refers to *sādhyasādhanabhāva* to explicate the relation between *liṅga* and *liṅgin*. He further explains this relation as 'having the same substratum' when he describes the function of *upanaya* as,

साधनभूतस्य धर्मस्य साध्येन धर्मेण सामानाधिकरण्योपपादनम् उपनयार्थः ।[15]

Here *sāmānādhikaraṇya*, which means having the same substratum, is the explanation of *sādhyasādhanabhāva*. This is how Vātsyāyana was thinking about the question as to how we could conceive of analogy which is strong enough to be called as *pramāṇa*. He thought that in *anumāna*, we must have seen that in all the observed cases, *liṅga* and *liṅgin* have the same substrata, that is to say, they must be either co-existent or co-absent. He also thought that while doing *anumāna*, we should not pass from one positive instance and/or one negative instance to the *sādhya*-case, but we must have observed a number of instances. That is why, he exemplifies *dṛṣṭānta* as *'sthālyādivat'*[16] (like the earthen pot and other things), *'yathātmādidravyam'*[17] (like the soul and the other substances) and so on.

§ 2.7

While explaining *hetu*-statement, *dṛṣṭānta*-statement and *upanaya*-statement, Vātsyāyana gives as example, some seemingly general statements like 'उत्पत्तिधर्मकमनित्यं दृष्टमिति ।'[18] 'अनुत्पत्तिधर्मकं नित्यं दृष्टम् ।'[19] The question about such statements is, however, that whether he takes these statements to be generally true of the known world, or universally true irrespective of the known or unknown world. The seemingly general statements which Vātsyāyana gives in his commentary on *NS* 1.1.38 are worth considering in this context. There he says,

"The substances like pot are seen to be created as well as impermanent".[20]

"The substances like soul are seen to be non-created as well as permanent".[21]

Such formations of his examples of 'general statements' sufficiently convey that he is taking these statments as the generalisations about the observed world. And this is important from the point of view of distinguishing between the inference based on 'universal concomitance' and the *anumāna* of Vātsyāyana. The statement that conveys the universal concomitance, if true of the non-empty universe, can be instanciated by all the observed as well as unobserved cases. The general statement in Vātsyāyana's frame of *anumāna* however, can be instanciated only by observed cases. For instance, one has to consider seriously Vātsyāyana's wording in, 'उत्पत्तिधर्मकमनित्यं दृष्टमिति ।'[22] The term *dṛṣṭam* indicates that here Vātsyāyana is concerned with the observed data. When Vātsyāyana gives such instances of general statements, he often uses the term *dṛṣṭam* to indicate the above idea. Vātsyāyana never uses the classical form of vyāpti, *'yatra yatra ..... tatra tatra ......' nor does he mention any universal statement containing the term synonymous to 'all'.*[23]

§ 2.8

It may be noted here that Vātsyāyana's notion of constant relation between *liṅga* and *liṅgin*, even in the observed world, was that of equivalence and not that of implication. Here the general statement namely 'अनुत्पत्तिधर्मकं नित्यं दृष्टम्' needs more explanation. DHRUVA while interpreting this, says, "Here the predicate is placed before the subject just as in *'anityaḥ śabdaḥ'*. Therefore it is नित्यमनुत्पत्तिधर्मकम् and is thus a good *vyatirekavyāpti*". (*NP*, Intro. p. xxviii, footnote).

But in fact the above-quoted statement of Vātsyāyana clearly shows that Vatsyāyana was not aware of the proper direction of the *hetu-sādhya*-relation, i.e. the *sādhya-sādhanabhāva*. His notion of the constant relation was ony that of co-existence and co-absence. It is worth noting here that implication does not have commutative property. Thus 'whatever is X is Y' cannot be equivalent with 'Whatever is Y is X'. But 'X is equivalent with Y' is equivalent with 'Y is equivalent with X'. The statement 'X is equivalent with Y' can be defined as '(X and Y) or (not-X and not-Y)' which clearly expresses the form of *sādhyasādhanabhāva* as 'co-existence or co-absence'. The same statement can also be defined as '(Y because of X) and (not-Y because of not-X)'. We also find this form indicated by Vātsyayāna.

For instance, while explaining positive instance, Vātsyāyana says, "Sound is impermanent because of being created, like the pot etc."[24] And while explaining negative instance, he says, "Here in case of the instances like soul, the impermanence does not exist because the createdness does not exist."[25] If Vātsyāyana had the notion of *vyāpti* as that of an implication, here he would have presented the transposed form namely, 'Here in case of the instances like soul, the createdness does not exist because the impermanence does not exist'. However, Vātsyāyana's wrong formulation clearly indicates that his concept of constant relation was that of observed equivalence.[26] A consequence of this is that both the cases in which *hetu* exists without *sādhya* and those in which *sādhya* exists without *hetu* are regarded as the cases of savyabhicāra.[27]

§ 2.9

Vātsyāyana follows the distinction between *anumāna* and *jāti*, made by Akṣapāda. While explaining the distinction Vātsyāyana says, "One who is using *jāti*, makes his counter-argument without establishing *sādhyasādhanabhāva* (probandum - probans-relation) observed in the instance. On the contrary, we can use probans as *hetu* (in the *anumāna*), only when the *sādhyasādhanabāva* among the two properties is established in the instance. We cannot use probans as *hetu*, which is mere similarity or difference."[28] While interpreting this we have to note that Vātsyāyana's notion of *sādhyasādhanabhāva* is not that of implication between *hetu* and *sādhya*, but that of 'observed equivalence', as distinguished from 'material equivalence'.[28]

While explaining the distinction between *anumāna* and *jāti*, Vātsyāyana distinguishes between *sāmānyahetu* and *viśeṣahetu*.[29] The notion of *sāmānyahetu* resembles the necessary condition of *sādhya* while the notion of *viśeṣahetu* resembles the necessary and sufficient condition of *sādhya*. But Vātsyayāna in his mind must have only the primary meanings of the terms *viśeṣa-hetu* and *sāmānyahetu*. By *viśeṣahetu* he meant the peculiar *hetu*, which and which alone proves the *sādhya* and by *sāmānya-hetu* he meant the general *hetu* which cannot conclusively prove *sādhya*. But the question what makes any *hetu* a *sāmānyahetu* or a *viśeṣahetu* was not answered by Vātsyāyana. For answering the question in the light of his concept of anumana, we have again to use his notions of *sādhyasādhanabhāva* and *sāmānādhikaraṇya*, which are suitable only to a form of strong analogy.

Part II
§ 2.10

According to Akṣapāda there are five kinds of *hetvābhāsas*, 1) *savyabhicāra* 2) *viruddha 3) prakaraṇasama* 4) *sādhyasama* 5) *kālātīa*.[30] All the Nyāya logicians following Akṣapāda accept this as a proper classification of *hetvābhāsas*.[31] In later treatises, beginning from Jayantabhaṭṭa's *Nyāyamañjarī*, these *hetvābhāsas* are treated as the violations of one of the five necessary conditions of sound *hetu*. According to these later logicians, the necessary conditions of sound hetu are strictly five and the corresponding violations of these conditions are also five. Thus according to them, there are only five *hetvābhāsas* primarily, though the number can be raised by subclassifications and combinations of these primary kinds. Such a systematisation of *hetvābhāsas* was probably due to the influence of Dinnāga and Dharmakīrti. But the Nyāya logicians earlier to Dinnāga, and even Udyotakara who appeared after Dinnāga, did not think in terms of the five and strictly five necessary conditions of sound *hetu*. Akṣapāda had simply made a list of the types of *hetus* which are unsound, but appear sound. These types were known to him from the actual debates that took place during his time. Vātsyāyana thinks that *hetvābhāsas* occur due to the violation of the defining characteristics of the sound *hetu*, but he does not make clear what according to him are those defining characteristics.

We shall consider below Akṣapāda's definitions of five *hetvābhāsas* and also Vātsyāyana's own account arising from his interpretations of Akṣapāda's aphorisms.

§ 2.11

We see that in *Carakasaṁhitā* the term *savyabhicāra* is used in a broader sense, in the sense of 'related irregularly or with uncertainty.'[33] Irregular relation between x and y means x is sometimes related with y but sometimes not. Uncertain relation between x and y means x might or might not be related with y. Sometimes the term *savyabicāra* applied to the relation between x and y means that x is not related with y at all. For example when Akṣapāda defines *pratyaksa* as a kind of *avyabhicāri* cognition, he is suggesting that when any particular cognition is said to be *savyabhicāra* in this sense, its content is supposed to be utterly unrelated with the object.[34] But when x is *savyabhicāra* with y in the first sense,[35] x is sometimes related with y and sometimes not. When Akṣapāda talks of *savyabhicāra* as a kind of *hetvābhāsa*, he uses the term in this sense. The *hetu* would be supposed to be *savyabhicāra* when it is observed to be related with *sādhya* in some places, but not related with *sādhya* in some other places. Therefore in the case of inferring such a *sādhya* from such a *hetu*, the *hetu* would not be the sufficient condition either of *sādhya* or of the absence of *sādhya*. It would not be possible for us in that case to reach either of the two opposite conclusions and thus the *hetu* would become *anaikāntika*. Thus Akṣapāda defines *savyabhicāra*, as *anaikāntika*, where *anaikāntika* means 'that which does not lead to any definite end (= conclusion)'.[36]

§ 2.12

We have seen that Vātsyāyana's notion of *sādhyasādhanabhāva* is that of 'observed equivalence'.[37] Here, while discussing his interpretation of *savyabhicāra* also, we can see that Vātsyāyana's notion of *savyabhicāra* was that of the observed negation of the so-called observed equivalence, rather than that of the negation of implication. Vātsyāyana defined *vyabhicāra* as *'ekatra avyavasthā'*[38] (the state of non-co-existence) which means the situation in which one relatum exists in absence of the other relatum. The situations described as

1) *Hetu* exists without *sādhya*

and 2) *Sādhya* exists without *hetu*.

are the forms of *vyabhicāra* according to Vātsyāyana.

The example of the inference involving *savyabhicāra*, given by Vātsyāyana, runs,

Sound is eternal

because it is intangible

like the soul and unlike the pot.

It is clear that from the classical point of view the above *hetu* ('being intangible') is a case of *sādhāraṇa-savyabhicāra*, because here the *vyabhicāra* that could be shown is 'cognition is intangible but not eternal'. And Vātsyāyana does mention this *vyabhicāra*. But he also mentions another case of *'vyabhicāra'* which from the classical point of view is not a case of *vyabhicāra* at all. The second *'vyabhicāra'* according to Vātsyāyana runs, 'atoms are tangible but eternal'. This certainly is not the negation of *vyāpti*-statement namely 'Whatever is intangible is eternal'.[39]

This attitude of Vātsyāyana, however, suggests that the concept of *vyabhicāra*, like the concept of *sādhyasādhanabhāva*, was not properly worked out by Vātsyāyana. Diṅnāga seems to have explained these concepts much better. According to Diṅnāga's understanding, it would not amount to *savyabhicāra* if some instances are found in which *sādhya* exists without *hetu*, but it would certainly amount to *savyabhicāra* if any instance is found where *hetu* exists without *sādhya*. Even Udyotakara who came after Diṅnāga was either not aware of this incorrect attitude of Vātsyāyana, or he ignored it. Vācaspatimiśra seems to have noticed this oddness in Vātsyāyana's interpretation of *savyabhicāra*, but he tried to protect Vātsyāyana by ascribing different intention to him. Vācaspatimiśra says, "Somebody may object that in case eternity is not the *sādhya*, but intangibility is the *sādhya*, to be proved by the *hetu*, namely eternity, (how then cognition, which is intangible but not eternal, present a case of *savyabhicara*?) Keeping this objection in mind, Vātsyāyana says that atom is tangible and also eternal etc."[40]

§ 2.13

We have seen that Caraka includes *viruddha* in the list of *vākyadoṣas*, that is, statement-faults.[41] It seems from the Akṣapāda's definition of *viruddha* that the *viruddha* of Akṣapāda is only a species of *viruddha* described by Caraka. Akṣapāda defines *viruddha* as,

सिद्धान्तमभ्युपेत्य तद्विरोधी विरुद्धः ।[42]

Meaning: If an arguer, after accepting some thesis, infers with the help of some *hetu* which is contrary to the accepted thesis, then the contrary *hetu* is called as *viruddha*.

Thus *viruddha* of Akṣāpāda can be considered as a type of faulty *hetu*-statement and can be compared with *siddhāntaviruddha* of Caraka.

It is clear that *viruddha-hetvābhāsa* according to Akṣapāda does not mean the *hetu* contrary to *pratijñā* or *sādhya* as was supposed later on by classical theoreticians. It was mainly under the influence of Vaiśeṣikas' and Diṅnāga's formulation of *viuddha* that the commentators of *Nyāyasūtra*, who appeared after Udyotakara, identified *viruddha* of Akṣapāda with the *viruddha* of Vaiśeṣikas and Buddhists. Thus the Nyāya-aphorism was misinterpreted by the later commentators like Jayanta so as to mean that *viruddha* is that *hetu* which is contrary to *sādhya* which is claimed to be true. Thus *siddhānta* was interpreted as *sādhya*[43] and *abhyupetya* was interpreted as *pratijñāya*[44] ('after claiming'). Both these meanings are not intended by Akṣapāda at all. In *Nyāyasūtra* the term *siddhānta* is nowhere used to mean *sādhya*. Similarly it is clearly seen that *abhyupetya* means 'after taking for granted'. Akṣapāda uses the term *abhyupagama-siddhānta* in which *abhyupagama* means 'accepting'[45] *Abhyupagama is derived from abhi* + *upa* + $\sqrt{gam}$ while *abhupetya* also is derived from a synonymous root *abhi* + *upa* + $\sqrt{i}$. Thus *'siddhāntam abhupetya'* means 'after accepting a thesis'.

§ 2.14

The example of *viruddha* given by Vātsyāyana is in tune with Akṣapāda's notion of *viruddha*. The example runs,

(The inference involving *viruddha-hetvābhāsa* -)

This transformation (in the thing) makes a deviation from (the identity of) the thing,

because permanence (of the thing) is denied.

(The thesis previously accepted by the debater -)

Even if the thing deviates (from its original pattern), it is (still the same), because the annihilation (of any thing) is denied.[46]

Here the *hetu* namely 'denial of permanence' is contrary to the accepted thesis that the changing thing remains the same (i.e. does not lose its identity). The thesis previously accepted by the arguer is external to the inference he is making now. In this sense this fallacy may be called the external fallacy of *viruddha*. This should be distinguished from the internal fallacy of *viruddha* which occurs due to the contrarity between *hetu* and *sādhya*. The *viruddha- hetvābhāsa* discussed by the later logicians was primarily the internal fallacy of *viruddha*.

Udyotakara's commentary on *NS* 1.2.6 gives too wide a notion of *viruddha*, which is applicable to any *hetvābhāsa*. This was obviously not meant by Akṣapāda.[47]

§ 2.15

We find that *prakaraṇasama* has been mentioned by caraka in his list of three *ahetus*.[48] The meaning of *prakaraṇasama* has not changed from Caraka to Akṣapāda.

Akṣapāda defines the *prakaraṇasama-hetvābhāsa* as,

यस्मात् प्रकरणचिन्ता स निर्णयार्थम् अपदिष्टः प्रकरणसमः ।[49]

Meaning - *Prakarṇasama* is one from which the investigation of the problem begins and still which is applied as a *hetu* for reaching the conclusion.

The term *prakaraṇa* has been left undefined by Akṣapāda. Etymologically *prakaraṇa* is that which is being done. In present context the problematic issue that is being discussed for solution may be called *prakaraṇa*. The act of investigating (the problem) may be called its *cintā*.

Udyotakara's interpretation of *prakaraṇasama* suggests the same thing. In his *Vārtika* he distinguishes *sādhyasama* from *prakaraṇasama*. In the case of *sādhyasama* (what he calls *sādhyāviśiṣṭa*) the given *hetu* is similar to (and not the same as) *sādhya* while in the case of *prakaraṇasama* the root of the issue itself is applied as the *hetu*.[50]

§ 2.16

RANDLE while explaining *prakaraṇasama* identifies it with petitio principii[51] (I think that he is correct in doing so). But MATILAL does not agree with this[52] and this has created a controversy. MATILAL disagrees on the count that the explanation of *prakaraṇasama* given by Vātsyāyana is clear enough to show that *prakaraṇasama* is the fallacy of indecision rather than petitio principii. MATILAL also distinguishes the fallacy of indecision *(prakaraṇasama)* from that of inconclusive reason *(savyabhicāra)* for the ways in which these fallacies are shown, differ. MATILAL however trusts in the Vātsyāyana's interpretation of *prakaraṇasama*. Yet another explanation may be suggested that Akṣapāda's definition of *prakaraṇasama* may be interpreted in the light of the account of *prakaraṇasama-ahetu* as given by Caraka, who is prior to Akṣapāda and whose account of logic and epistemology is similar to that of Akṣapāda in content and style. Secondly, the definition of *prakaraṇasama-hetvābhāsa* given by Akṣapāda, if taken literally, is more suggestive of petitio principii than of any other fallacy. Akṣapāda in his definition asserts the identity between the problematic issue and the applied *hetu*, as the condition of the fallacy.

§ 2.17

Akṣapāda has employed the same word *prakaraṇasama* to denote a kind of *jāti*[53] (=sophistry based on similarly or difference). His definition of *prakaraṇasama* as a kind of *jāti* is different from his definition of *prakaraṇasama* as a *hetvābhāsa*. It is remarkable that the *prakaraṇasama-jāti* of Akṣapāda resembles the fallacy called *satpratipakṣa-hetvābhāsa* of classical Nyāya logicians. It is possible that Vātsyāyana was guided by the Akṣapāda's definition of *prakaraṇasama-jāti* and almost identified it with *prakaraṇasama-hetvābhāsa*.

Vātsyāyana defined *prakaraṇa* as the two opposite views which raise a doubt and which are not settled (by deciding the truth of one of them).[54] By *cintā* of the *prakaraṇa*, Vātsyāyana meant the investigation of the problem.[56] However, this is not the only sense given to *prakaraṇasama* by Vātsyāyana. He adds that this fallacious *hetu* appears to prove both the opposite views and thus cannot overcome the *prakaraṇa*.[57] Thus *prakaraṇasama-hetvābhāsa* as described by Vātsyāyana resembles more with the *satpratipakṣa-hetvābhāsa* of later Nyāya, than with petitio principii.

The discussion shows that Vātsyāyana certainly deviated from the original meaning of *prakaraṇasama* and created the ground for later logicians to consider *satpratipakṣa* as a *hetvābhāsa*. The Nyāya authors after Udyotakara almost took it for granted that by *prakaraṇasama* Akṣapāda meant *satpratipakṣa*.

§ 2.18

We do not find any mention of *sādhyasama* in *Carakasaṁhitā*. *Saṁśayasama* mentioned by Caraka is based upon the dubitability of the relation between *hetu* and *sādhya* and not on the dubitability of the existence of *hetu* in *pakṣa*.[58] Thus *sādhyasama* was probably the original contribution of Akṣapāda.

The definition of *sādhyasama* given by Akṣapāda runs,

साध्याविशिष्टः साध्यत्वात् साध्यसमः ।[59]

Meaning - *Sādhyasama* is that (fallacious *hetu*) which is not distinguished from *sādhya* because that itself needs to be proven.

In the above definition the word *sādhya* is used in two different ways. In '*sādhyāviśiṣṭa*' *sādhya* is used as a technical term and means the probandum. But in *sādhyatvāt*, '*sādhya*' is used in its etymological sense and means 'that which has got to be proven'. Vātsyāyana recognised this when he said, 'Because this kind of fallacious *hetu* is not proved to be true of the relevant case, it has got to be proved like *sādhya*'.[60]

The example given by Vātsyāyana is as follows,

Shadow is a substance

because it moves.

Here the movement subsisting in shadow is the fallacious *hetu*, because it is dubitable whether movement subsists in a shadow in the sense in which it can subsist in a substance. Thus in other words, the subsistance of *hetu* in *pakṣa* being dubitable, is insufficient to prove the subsistance of *sādhya* in it.

§ 2.19

Asiddha-hetvābhāsa of the classical Nyāya resembles *sādhyasama*. But all the classical subtypes of *asiddha* were not known to Akṣapāda or Vātsyāyana. The *Sādhyasama* of Akṣapāda and Vātsyāyana more or less resembles *svarūpāsiddha* of classical Nyāya. But there lies certain difference even between *sādhyasama* and *svarūpāsiddha*. We generally find in the classical examples of *svarūpāsiddha* that the *hetus* used in those examples are not only not proved as being present in *pakṣa*, but they are also clearly absent in *pakṣa*. Thus the term *asiddha* means 'proved negatively' or 'proved absent' and not just 'unproved'. In the classical example of *svarūpāsiddha* this is clearly seen.

'The sound is permanent

because it is perceptible through eyes'.

Here it is not the case that 'the perceptibility of sound through eyes' is not proved. It is rather disproved by our experience. The *sādhyasama-hetvābhāsa* of Akṣapāda and Vātsyāyana is not a disproved *hetu*. Thus it seems that one variety of *svarūpāsiddha* which is called *sandigdhāsiddha*[61] by Dignāga and Dharmakīrti, can be compared with *sādhyasama*.

§ 2.20

RANDLE, identifies *sādhyasama* with petitio principii.[62] MATILAL on the other hand states that in the case of petitio principii the reason *(hetu)* can be shown to be either the same as or logically dependent upon the conclusion. This is not necessarily true of *sādhyasama.*[63] According to MATILAL *sādhyasama* is the fallacy of the same predicament.

We also find some ambiguity in the use of the term *sādhyasama-hetvābhāsa* in ancient times. Gauḍapādācārya used the term to connote *sādhyasama dṛṣṭānta.*[64]

§ 2.21

Akṣapāda's definition of *kālātīta-hetvābhāsa* is not quite clear and Vātsyāyana's explanation also is confusing. It can be, however, pointed out that the classical interpretation of *kālātīta* as *bādhita* is not plausible. Vātsyāyana's interpretation of *kālātīka* also supports this thesis. *Aprāptakāla* considered by Akṣapāda in the list of *nigrahasthānas* (the occasions of defeat) is clearly not identical with *kālātīta*. The occasion of defeat *aprāptakāla* occurs when the *hetu* in the five- membered inferencial statement, is not presented in proper temporal order, i.e., when its conventional order is altered.[64a] It is not the intention of Akṣapāda that in the case of *aprāptakāla*, *hetu* itself is improper, i.e., fallacious. On the other hand, if any *hetu* is of *kālātīta*-kind, then that *hetu* is fallacious.

Caraka has mentioned *kālātītavacana*, which is an occasion of defeat. By *kālātītavacana* he means, saying something in a debate which had to be said earlier in the debate, or confuting the other debater when the proper time of confutation has gone by.[65] *Aprāptakāla* of Akṣapāda thus resembles *kālātītavacana* because both are concerned with the time of presentation of the statements in the course of a debate. Neither of the two is concerned with the soundness or unsoundness of the *hetu* presented. This difference between *kālātīta-hetvābhāsa* and *aprāptakāla* as an occasion of defeat has been noticed by Vātsyāyana and accepted by all the later commentators.

The question remains, however, what *kālātīta- hetvābhāsa* means. The definition of *kālātīta* in *Nyāyasūtra* is

कालात्ययापदिष्टः कालातीत : ।[66]

Literally the aphorism means that *kālātīta* is one which is provided as *hetu* when the time lapses. It is not made clear as to which time Akṣapāda is speaking about. However, we have to make a guess. As *kālātīta* is a *hetvābhāsa*, i.e., a fallacious *hetu*, the fallacy of *hetu* does not consist in its being presented a little late. Thus the lapse in time must be with reference to the thing or event itself which is being communicated as *hetu* and not with reference to the communication of *hetu* . The aphorism then means: When something (i.e., a thing or a property or an event) is applied as a *hetu* with a lapse with reference to its (proper) time (of being connected with *sādhya*) then that (pseudo-)*hetu* is called *kālātīta*.

§ 2.22

This point may be clarified by an example. Take a case when somebody is inferring the occurrence of some cause-event with the help of the *hetu* which is an effect-event. The cause-event should be necessarily prior to the effect- event. If the effect occurs at time t_2, the cause must occur at some moment prior to time t_2. We call this prior moment as t_1. If, while inferring the cause from the effect, this gap between between t_1 and t_2 is ignored, the argument may become fallacious. The popular inference with 'fire on the mountain' as *sādhya* and 'smoke on the mountain' as *hetu*, if made by ignoring the time- gap, becomes fallacious. Let us state the inference,

The mountian is smoky

Whatever is smoky is fiery

Therefore the mountain is fiery.

Here, if the second premise is taken to connote the causal connection between fire and smoke we should state the premise as 'whatever is smoky at sometime must be fiery at the time preceding it'. Accordingly if the first premise is in present tense '(The mountain *is* smoky'), then the conclusion of the causal inference should be in past tense ('Therefore the mountain was fiery').

In Indian logic such an example is never given as an example of *kālātīta*. But if Akṣapāda's frame of *kālātīta* is to be applied strictly, then this example has to be regarded as a case of *kālātīta*.[67]

§ 2.23

Another variety of *kālātīta* is seen in an inference where the temporal gap between the time of being of something and the time of the appearance or apprehension of that, is ignored and confused. In the case of visual appearance, for instance, there is a gap between the time when the thing is present before the spectator's eyes and the time when the same is visually sensed. The thing which is present before us is not seen by us at the very moment. This is so because the rays of light take a little time for passing from the thing to our eyes. The gap is, however, very small and we prefer to neglect it. Similar temporal gap can be observed in the case of the auditary appearance of some sound produced. The temporal gap between the occurrence of a sound and the auditary sensation of the sound is far bigger than the gap in case of a visual appearance. The Vātsyāyana's example of *kālātīta* is based upon a similar understanding.

The example runs,

The sound is permanent because it is manifested by the contact (of things), like colour (which is manifested by the light thrown upon it by the lamp).[68]

In the above example, Vātsyāyana imagines a situation where a person is hitting a log of wood with an axe and another person hears the sound of hitting from a long distance. The other person does not hear the sound at the very moment when the axe and the log come in contact with each other, but he hears it when the contact is over.[69] Thus according to Vātsyāyana the appearance of sound is not the manifestation of sound. The colour of a pot, on the contrary, appears at the very moment when the contact between the light and the pot occurs.[70] Thus according to Vātsyāyana, the appearance of a colour can be regarded as the manifestation of the colour. (Vātsyāyana, however, did not know that even in case of colour there is a temporal distance between its contact with light and its perception by a perceiver, though the distance is very small as compared with one in the case of sound). Vātsyāyana thought that if the sound is manifested only through the contact between the axe and the log, it should appear at once just as the colour appears. Thus one who argues that sound is just manifested and not originated through the contact of things, ignores the temporal gap between the occurrence and appearance of sound and thus commits the fallacy called *kālātīta*.

Thus *kālātīta* is a fallacy which is committed by confusing the time-factor involved in the argument with reference to the occurrence and appearance of *hetu*, *pakṣa* or *sādhya*, or with reference to the relation between them.

Udyotakara is in tune with Vātsyāyana when he distinguished between *upalabdhikāla* (the time of appearing) and *saṁyogakāla* (the time of contact).[71]

Notes

1. The words *pramāṇa*, *upamāna* and *anumāna* in the sense of the kinds of cognition, are derived by adding *'bhāve lyuṭ'* to *pra* + √*mā*, *upa* + √*mā* and *anu* + √*mā* respectively.
2. इन्द्रियार्थसन्निकर्षोत्पन्नं ज्ञानमव्यपदेश्यमव्यभिचारि व्यवसायात्मकं प्रत्यक्षम् । (NS 1.1.4). Jayantabhaṭṭa truly thinks that the terms avyabhicāri and vyavasāyātmaka (and also jñāna) are the definians applicable to pramāṇa in general. See NMr, Part I, p. 113.
3. It is possible to remove the difficulty only if we explain the logical connective 'or' in terms of the Navya Nyāya concepts of *anyatara* and *anyatama*, rather than in terms of doubt *(saṁśaya)*. However the question remains how *saṁśaya* can be formally explained.
4. If *avyabhicāri* is taken in such a general sense, *avyabhicāra*, in the context of inference, becomes synonymous to *avinābhāva*. Though the popular meaning of *avinābhāva*, like that of *avyabhicāra*, is *vyāpti*, in classical theories of logic we find that all the three (according to Buddhists) or all the five (according to Naiyāyikas)

constitutive elements of sound *hetu* were covered by the notion of *avinābhāva*. Jayantabhaṭṭa, for instance, opines, सोऽयमेतेषु पञ्चसु लक्षणेषु. अविनाभावो लिङ्गस्य परिसमाप्यते । *NMr*, Part I, p. 101

5. I have borrowed the expressions 'signifying analogy' and 'signified analogy' from COHEN and NAGEL. See *ILSM* p. 287. These terms aptly denote *hetu* and *sādhya* respectively, in the context of the theory of *anumāna* as analogical reasoning.
6. I have borrowed the expressions 'positive analogy' and 'negative analogy' from KEYNES. See *TP*, Chap. XIX, p. 223. These terms aptly denote *sādharmya* and *vaidharmya* respectively.
7. Compare with Caraka who defines *anumāna* as *tarka*. See 1.5.
8. Compare with Caraka. See 1.5.
9. तत्पूर्वकं त्रिविधमनुमानं पूर्ववत्, शेषवद्, सामान्यतो दृष्टं च । *NS* 1.1.5.
10. उदाहरणसाधर्म्यात् साध्यसाधनं हेतुः । *NS* 1.1.34; तथा वैधर्म्यात् । *NS* 1.1.35
11. साध्यसाधर्म्यात् तद्धर्मभावी दृष्टान्त उदाहरणम् । *NS* 1.1.36; तद्विपर्ययाद् वा विपरीतम् । *NS* 1.1.37.
12. उदाहरणापेक्षस्तथेत्युपसंहारो न तथेति वा साध्यस्योपनयः । *NS* 1.1.38.
13. Though DHRUVA's observation in case of *dṛṣṭānta*, namely, '.... the fact that the word *yathā* is used instead of *iva* is significant' (NP, Intro., p. xxx) is correct, still the distinction between *yathā* and *iva* can be drawn as parallel to the distinction between analogy and apparent (may be, poetic) similarity.
14. Thus DHRUVA's supposition namely, "Another word used in *Nyāyasūtras* which bears clear testimony to the recognition of the general law as the basis of Inference is *'savyabhicāra'* -i.e. the *hetu* with *vyabhicāra* or variability as opposed to invariability which is its necessary implication as a condition of valid reasoning" (*NP*, Intro, p. xxviii) seems to be wrong.
15. *NBh* on *NS* 1.1.39.
16. *NBh* on *NS* 1.1.36.
17. *NBh* on *NS* 1.1.35.
18. *NBh* on *NS* 1.1.34.
19. *NBh* on *NS* 1.1.38.
20. स्थाल्यादिद्रव्यम् उत्पत्तिधर्मकम् अनित्यं दृष्टम् । *NBh* on *NS* 1.1.38.
21. आत्मादिद्रव्यम् अनुत्पत्तिधर्मकं नित्यं दृष्टम् । *NBh* on *NS* 1.1.38.
22. *NBh* on *NS* 1.1.34.
23. UI refers to Piṅgala, a commentator of *Mādhyamika Śāstra*, who arose before the time of Vātsyāyana. (See *VP*, p. 87) Piṅgala describes the inference of fire from smoke. Pingala says "If a man has once seen that fire has smoke; afterwards he can infer the

existence of fire from seeing smoke". UI comments that the first part of the description is not strictly logical. However, we have to note here that Piṅgala's understanding of inference suggested by the above quotation is in tune with the general understanding of *anumāna* before Vātsyāyana.

24. शब्दोऽप्युत्पत्तिधर्मकत्वाद् अनित्यः स्थाल्यादिवद् इति । *NBh* on *NS* 1.1.36.

25. अत्रात्मादौ दृष्टान्ते उत्पत्तिधर्मकत्वस्याभावाद् अनित्यत्वं न भवति । *NBh* on *NS* 1.1.37.

26. For Symbolic representation see Appendix III. KEITH seems to have noticed this when he says with reference to the treatises of Diṅnāga and Udyotakara, ".... it is expressly recognised that the extension of the middle is not equivalent to that of the major; fire may exist without smoke. This point the earlier formulation ignored, for the simple reason that it was irrelevant to the business in hand ....." *ILA*, p. 140.

27. This will be clear again when we come to Vātsyāyna's interpretation of *svyabhicāra*. See § 2.12.

28. अव्यवस्थाप्य खलु धर्मयोः साध्यसाधनभावम् उदाहरणे जातिवादी प्रत्यवतिष्ठते । व्यवस्थिते तु खलु धर्मयोः साध्यसाधनभावे दृष्टान्तस्थे गृह्यमाणे साधनभूतस्य धर्मस्य हेतुत्वेनोपादानं न साधर्म्यमात्रस्य न वैधर्म्यमात्रस्य वेति । *NBh* on *NS* 1.1.39., DHRUVA refers to this. See *NP*, Intro., p. xxix.

29. See, for instance, *NBh*, on *NS* 5.1.2.

30. See *NS* 1.2.4.

31. With the exception of Bhāsarvajña who accepted six basic types of *hetvābhāsa*. See § 6.27.

32. हेतुलक्षणाभावाद् अहेतवो हेतुसामान्याद् हेतुवदाभासमानाः ..... *NBh* on *NS*. 1.2.3.

33. See § 1.7.

34. This sense suits to the Śaṅkara's notion of *adhyāsa*. Cf. यत्र यदध्यासः तत्कृतेन दोषेण गुणेन वा अणुमात्रेणापि स न सम्बध्यते । *BSSB*, *Adhyāsabhāṣya*. Any particular false cognition can be called as *savyabhicāra* in this sense. See § 2.1.

35. Any particular cognition cannot be called as *savyabhicāra* in this sense, because one and the same cognition cannot be 'sometimes true and sometimes false'.

36. अनैकान्तिकः सव्यभिचारः । *NS* 1.2.5. KEITH translates *anaikāntika* as 'which leads to more conclusions than one' (See *ILA*, p. 131) which is obviously wrong. *Anāikāntika* is just the negation of *aikāntika* (= which leads to a definite conclusion), but in the sense attached to the term by KEITH, the apt term would have been *ānekāntika*, which would be derived from *'anekānta'* (=more conclusions than one).

37. See § 2.8. I have coined the expression 'observed equivalence' to connote the equivalence between *hetu* and *sādhya*, noticed in the observed world. Thus observed equivalence is distinct from the material equivalence (or the strict implication as BARLINGAY calls it. See *MIIL*, pp. 165-6).
38. See *NBh* on *NS* 1.2.5.
39. RANDLE has noticed this. See *ILES*, p. 193, fn. 1.
40. ननु च मा भून्नित्यत्वं साध्यम् अस्पर्शत्वम् एवास्तु नित्यत्वस्य साध्यम् इत्यत आह, स्पर्शवांश्चाणुर्नित्यश्चेति । *NVT* on *NS* 1.2.5.
41. See § 1.7.
42. *NS* 1.2.6.
43. See *NMr*, Part II, pp. 156-7.
44. See *NMr*, Part II, p. 158.
45. See *NBh* on *NS* 1.1.31.
46. See *NBh* on *NS* 1.2.6.
47. नन्वयम् एको हेत्वाभासो विरुद्धः प्राप्नोति । सत्यम् एक एव हेत्वाभासो विरुद्धः । *NV* on *NS* 1.2.6.
48. See § 1.6.
49. *NS* 1.2.7.
50. यत्र साध्येन समानो धर्मः हेतुत्वेन उपादीयते स साध्याविशिष्टः । यत्र पुनः प्रकरणप्रवृत्ति.हेतुरेव प्रकरणसमः । *NV* on *NS* 1.2.7.
51. See *ILES*, p. 196.
52. See MATILAL's article 'A Note on Nyāya Fallacy *Sādhyasama* and Petitio Principii', published in JIP, Vol. 2, Nos. 3-4 (March-June 1974).
53. See § 7.21.
54. विमर्शाधिष्ठानौ पक्षप्रतिपक्षौ अनवसितौ प्रकरणम् । *NBh* on *NS* 1.2.7.
55. तस्य चिन्ता विमर्शात्प्रभृति प्राक् निर्णयाद् यत् समीक्षणम् । *Ibid.*
56. सा जिज्ञासा यत्कृता स निर्णयार्थं प्रयुक्तः । *Ibid.*
57. उभयपक्षसाम्यात् प्रकरणमनिवर्तमानः प्रकरणसमः । *Ibid.*
58. See § 1.6.
59. *NS* 1.2.8.
60. अयमप्यसिद्धत्वात् साध्यवत् प्रज्ञापयितव्यः । *NBh* on *NS* 1.2.8.
61. For *sandigdhāsiddha* see §§ 3.23, 5.16.
62. See *ILES*, p. 197.
63. See MATILAL, op. cit.
64. बीजाङ्कुराख्यो दृष्टान्तः सदा साध्यसमो हि सः ।
    न हि साध्यसमो हेतुः सिद्धौ साध्यस्य युज्यते ।। *GK* 4.20.

64a. See § 7.44.

65. अतीतकालं नाम यत् पूर्वं तत् पश्चादुच्यते ...... पूर्वं वा निग्रहप्राप्तम् अनिगृह्य ..... पश्चान्निगृह्यते ..... । *CS*, p. 761.

66. *NS* 1.2.9.

67. Here the classical Nyaya philosophers may justify that they do not treat this inference as a causal inference. They may think that the general statement is made on the basis of co-existence *(anvaya)* and co-absence *(vyatireka)* without taking into account the temporal gap. This, I think, is not a sufficient justification because we can observe a situation where the fire has blown out just a moment before, but the smoke created by it is seen in that place. Such a situation would provide *vyabhicāra* which falsifies the general statement. Thus the temporal distance is to be mentioned while making such inference.

68. नित्यः शब्दः संयोगव्यङ्ग्यत्वाद् रूपवत् । *NBh* on *NS* 1.2.9.

69. निवृत्ते दारुपरशुसंयोगे दूरस्थेन शब्दः श्रूयते । *NBh* on *NS* 1.2.9.

70. सति प्रदीपघटसंयोगे रूपस्य ग्रहणं भवति । न निवृत्ते संयोगे रूपं गृह्यते । *Ibid.*

71. See *NV* on *NS* 1.2.9.

Chapter III

The Nature of Anumāna and Sādhanābhāsa According to Diṅnāga

Part I

§ 3.1

It is possible to argue that Diṅnāga and the Buddhist logicians like Vasubandhu before Diṅnāga, must have got some insight from *Vaiśeṣikasūtra*[1] in giving the account of three basic *hetvābhāsas*. But it has to be accepted that the account of *Vaiśeṣikasūtra* regarding *hetvābhāsas* was very crude, vague and narrow. It was so vague that only the misunderstanding of Kaṇāda's account could have helped Diṅnāga in the better understanding of good and bad *hetu*.[2] Diṅnāga was certainly responsible for the clear theorisation and detailed channelisation of the conditions of good and bad *hetu*. Though it is true that in some pre-Diṅnāga Buddhist texts on logic[3] we find the mention of the idea that the sound *hetu* should fulfil the three conditions namely *pakṣasattva, sapakṣasattva* and *vipakṣāsattva*, Diṅnāga seems to be first logician to work out the theory of reasoning based on it. Again, Diṅnāga seems to be the first logician to distinguish clearly between inference as a cognition accuring to oneself and inference as a communicative act projected to other person. The distinction is spelled out by him in the terms *svārthānumāna* and *parārthānumāna*.

Svārthānumāna as defined in *Nyāyapraveśa*[4] is the cognition of *sādhya* that occurs to a person who sees a *hetu*.[5] This cognition of *sādhya* is an intuiton *(darśana)* generated by *hetu*. The reliability of such intuition is determined by examining whether the *hetu* which generates the intuition, possesses the three constitutive elements of *hetu* or not. The inferring agent may not be, and generally is not, aware of each of the constitutive elements of *hetu*, when he actually intuits *sādhya*. But in *parārthānumāna* which is the communicative act of proving, the inferring agent tries to show that the *hetu* which has generated this intuition of *sādhya*, possesses all the three constitutive elements of good *hetu*. In this sense, it may also be noted that *svārthānumāna*, according to Diṅnāga, is not isomorphic with *parārthānumāna*.

§ 3.2

The three 'constitutive elements' (Heareafter, CEs) of *hetu* according to Diṅnāga are as follows,

1) The *hetu* must be present in the *pakṣa*-case

2) Out of the remaining cases, the *hetu* must be present in *sapakṣa*-cases.

3) The *hetu* must be absent in each of the *vipakṣa*-cases.

Here the term *pakṣa* is not used in the sense in which it was later on used by Naiyāyikas (that is, in the sense of the substratum, the presence of *sādhya* in which is doubted.). Diṅnāga meant by *pakṣa*, the thesis or the case which is being proved.[6] So the first CE of *hetu* states that the *hetu* must be true in the case of the thesis. In other words, *hetu* must be predicated of the subject of *pakṣa*. By analysing this first CE we get to assertions.

A1: There is the subject of the thesis.

(We call this subject as *pakṣa*-subject).

A2: *Hetu* is predicated of the *pakṣa*-subject.

§ 3.3

Sapakṣa means similar to *pakṣa*, that is to say, all those other cases which are similar to the thesis.[6a] In other words, *sapakṣa* means the cases in which the thesis comes out to be true. When one proves a thesis, he proves that the sādhya-predicate is predicated of the *pakṣa*-subject. Thus *sapakṣa* is similar to *pakṣa* in the sense that the *sādhya*-predicate, which is being proved to be predicated of the *pakṣa*-subject, is known to be predicated of the subject of the *sapakṣa*. So the second CE states that the *hetu* should be true of at least some of the *sapakṣa*-cases. The second CE also suggests that out of *sapakṣa*-subjects and *vipakṣa*-subjects, *hetu* is predicated only of *sapakṣa*- subjects. Though the term *sapakṣa- sattva* does not imply that *hetu* is predicated only[7] of *sapakṣa*-subjects and not of *vipakṣa*-subjects, it seems, however, that Dinnāga did have this import of *sapakṣa-sattva* in his mind, because according to him the correct presentation of this CE was of the form, 'That which has *hetu* is seen to have *sādhya*' and its converse namely 'That which has *sādhya* is seen to have *hetu*' was, according to him, fallacious due to *Viparītānvaya*.[7a] Thus by analysing the second CE we get two assertions.

B1 : There is at least one *sapakṣa*-subject of which *hetu* is predicated.

(Here by *sapakṣa*-subject we mean a subject which is not the *pakṣa*-subject, and of which *sādhya*-predicate is predicated).

B2 : *Hetu* is predicated only of *sapakṣa*-subjects (out of the subjects other than *pakṣa*-subject).

§ 3.4

Vipakṣa means the cases which are dissimilar to *pakṣa*. In other words, *vipakṣa* means the cases in which the thesis does not come out to be true. *Vipakṣa* is dissimilar to *pakaṣa* in the sense that *sādhya*-predicate which is being proved to be predicated of the *pakṣa*-subject, is known to be not predicated of the *vipakṣa*-subjects. Thus by analysing the third CE we get the two assertions,

C1 : There is at least one *vipakṣa*-subject of which *hetu* is not predicated.

(Here by *vipakṣa*-subject we mean the subject which is other than the *pakṣa*-subject and of which *sādhya*-predicate is not predicated).

C2 : *Hetu* is not predicated of any of the *vipakṣa*-subjects.

§ 3.5

Dinnāga must have thought in terms of some regular relation btween *hetu* and *sādhya* which is seen in the observed world, when he talked of *anvaya* and *vyatireka*. But there does not seem to be any satisfactory evidence to show that Dinnāga must have thought in terms of necessary connection[8] when he mentioned the three CEs of hetu. Dharmakīrti argued in *Hetubindu* that the word *niścita*[9] used by Dinnāga indicates the necessary relation which is expected between *hetu* and *sādhya*. But the word *niścita* used by Dinnāga is not exclusively indicative of necessity. It is also worth-nothing that Dharmakīrti found in Dinnāga's literature no better evidence for showing the concept of necessary connection than the ambiguous word *niścita*[9a]. *Niścita* could mean 'certain' and here it is perfectly sound to interpret Dinnāga as saying that *hetu* which certainly possesses three CEs, may not be necessarily related with *sādhya*. The term *'niyama'* which meant either necessity or the necessary relation was later on insisted upon by Dharmakīrti. Dinnāga, however, is not seen to use this word in this context.

It has been pointed out by STCHERBATSKY[10] and others, following Dharmottara that the second and the third CE imply each other, and thus, if examined strictly, the second (or third) CE is redundant. Dharmottara supposed that the intention behind mentioning the second and the third CE of *hetu*, when one of them is redundant, was to suggest some kind of necessity involved in the relation between *hetu* and *sādhya*.[11] Dharmakīrti on the other hand rightly held that 'existence in case of *sapakṣa*' and 'non-existence in case of *vipakṣa*' do not imply each other.[12] Even if we accept Dharmottara's opinion that second and third CEs imply each other, still the repetition of second CE in terms of third CE cannot imply the notion of necessity as Dharmottara supposed.

§ 3.6

The analysis of the three CEs made in 3.2 - 3.4 can help us in many ways. It can be seen from the analysis that neither second CE implies the thrid CE nor vice-versa. The assertion B_1 which is involved in the second CE is independent of the third CE. Similarly the assertion C_1 of the third CE is independent of the second CE. Thus the separate import of the second CE as well as of the third CE can be shown.

The analysis also suggests that though Diṅnāga talks of three CEs of *hetu*, the three CEs consist of many more assertions which result in the multiplication of the necessary conditions of *hetu*. This multiplication is automatically reflected in the multiplication of *hetvābhāsas*. This is why we can account for several sub-varieties of *asiddha*, *viruddha* and *anaikāntika hetvābhāsas*.

It can also be seen that all the assertions namely A_1, A_2, B_1, B_2, C_1 and C_2 taken together do not provide the logically sufficient ground for the proof of the thesis.

Let us consider the stock-example of inference in which sound is proved to be impermanent on the basis of its being a product. Sound in that case will be the subject of the thesis and impermanence will be the predicate, i.e. the *sādhdya*-predicate. That is to say, here we are proving impermanence to be predicated of sound. The characteristic of being a product is the *hetu*- predicate.

One who makes a *parārthānumāna*, first asserts the thesis and then for proving the thesis, shows that the *hetu* which is used to prove the thesis possesses all the three CEs. Therefore the total argument will look like the following :

Thesis: *Sādhya*-predicate is predicated of the thesis-subject.

Because,

First step: Hetu possesses first CE.

Second step: Hetu possesses second CE.

Third step: Hetu possesses third CE.

§ 3.8

But we have seen in some of the last sections that assertion of first CE amounts to assertion of A_1, and A_2, assertion of second CE amounts to assertion of B_1 and B_2 and assertion of third CE amounts to assertion of C_1 and C_2. Thus in the case of the example of inference referred to in the last section, the logical arrangement of the argument will be :

Thesis: Sound is impermanet.

Because,

First step: Sound exists (A_1) and has the characteristics of being a product (A_2).

Second step: There is at least one thing (other than sound) which is a product (B_1) and anything (other than sound) which is a product is impermanent (B_2).

Third step: There is at least one thing (other than sound) which is not impermanent (C_1) and anything (other than sound) which is not impermanent is not a product (C_2).

COMMENTS

(A) Since the statement that 'Sound is a product' presupposes or implies that sound exists, A_1 and A_2 need not be stated separately. So the first step could be simply 'sound is a product.'

(B) The second step can be shown to imply another statement that there is at least one thing (other than sound) which is both a product and impermanent. Instead of such an existential general statement, Diṅnāga expects that an instanciation of such a statement may be presented to replace the expression of B_1 in the second step. In the present example the instanciation can be stated as follows :

"Pot (which is obviously other than sound) is both a product and impermanent." We call this 'Revised B_1.'

Thus the second step may be stated as follows:

Anything (other than sound) which is a product is impermanent (B_2). For example, pot (which is obviously other than sound) is both a product and impermanent (Revised B_1).

(C) Similarly the third step can be shown to imply another statement that there is at least one thing (other than sound) which is neither a product nor impermanent. Instead of such an existential general statement, Diṅnāga expects that an instanciation of such a statement may be presented in the third step. In the present example the instanciation can be stated as follows ;

"Ether *(ākāśa)* (which is abviously other than sound) is neither a product nor impermanent." We call this 'Revised C_1.'

Thus the third step may be stated as follows :

Anything (other than sound) which is not impermanent is not a product (C_2). For example ether (which is obviously other than sound) is neither a product nor impermanent. (Revised C_1).

Thus the complete inferencial statement according to Diṅnāga will be like this :

Thesis: Sound is impermanent.

Because,

First step: Sound is a product

Second step: Anything (other than sound) which is a product is impermanent. For example pot (which is obviously other than sound) is both a product and impermanent.

Third step: Anything (other than sound) which is not impermanent is not a product. For example ether (which is obviously other than sound) is neither a product, nor impermanent.

We call the form of this *parārthānumāna* as P_1.

Although Diṅnāga's general frame of inference does not appear to allow *kevalānvayi*[13] to be a type of reliable *hetu*, Diṅnāga feels that we need not consider the assertion C_1 seriously in case of *kevalānvayi* type of *hetu*. In such a case, according to him, we can assert C_2 even though we reject the existence of *vipakṣa*.[14] The third CE namely *vipakṣāsattva* can be possessed by *hetu* according to Diṅnāga, either if there is no *vipakṣa* (and automatically no real negative instance) or if there is *vipakṣa* but *hetu* is not predicated of any *vipakṣa*- subject. The first possibility is true in case of *kevalānvayi* hetu. This is suggested by Diṅnaga when he says, ".....if that (negative instance) does not exist, (it is evident that the predicate) does not reside in it; then, since there is no possibility of doubt, such a fallacy (as pointed out) is non-existent".[15]

According to Diṅnāga, the general frame of *parārthānumāna* assumes the mention of the postive instance as well as the mention of the negative instance. The inference which cannot have positive instance was, according to Diṅnāga, a fallacy called *asādhāraṇa anaikāntika*. The negative instance according to him, however, need not necessarily refer to a case which actually exists; but the mention of negative instance was necessary while inferring, for showing a logically possible case where *hetu* and *sādhya* can be co-absent.[16]

This can be clear in the example we discussed in the last section. There the negative instance is that of ether which is stated to be both permanent and a non-product. But whereas a Naiyāyika while making such an inference can assert the truth of such a negative instance, a Buddhist logician will not accept it, because for him nothing is permanent. The example of ether for him may express at the most a logical possibility but not a material truth.

Thus in the case of the inference from *kevalānvayi* hetu, though the linguistic form of the argument is the same, the inner logical structure of the argument is different. The logical form of such an argument should be in fact presented by omitting the part of negative instance. Dinnāga's stock-inference in that case will be stated as follows :

Thesis: Sound is impermanet.

Because,

First step: Sound is a product.

Second step: Anything (other than sound) which is a product is impermanent. For example pot (which is obviously other than sound) is product and impermanent.

Third step: Anything (other than sound), if it is not impermanent then it is not a product.

(Here we have expressed third step as a conditional statement to avoid the sense of existential import.) We call the form of the above inference[16a] as P_2

§ 3.10

It is clear that both the forms of *parāthānumāna* namely P_1 and P_2 are invalid because in both of them the conclusion does not follow logically from the premises. However these forms of argument can be called the forms of strong analogical argument. In these, *hetu* stands for the positive signifying analogy between *pakṣa* and at least a part of *sapakṣa*. The same *hetu* stands for the negative signifying analogy between *pakṣa* and *vipakṣa*. In P_2, *hetu* does not stand for the negative signifying analogy because there *vipakaṣa* is absent. Thus hetu in P_2 may be called purely positive signifying analogy.[17] In both the arguments *sādhya* stands for the positive signified analogy between *pakṣa* and *sapakṣa*. In P_1 the same *sādhya* stands for the negative signified analogy between *pakṣa* and *vipakṣa*. In P_2, however, *sādhya* stands for purely positive signified analogy.

The difference between Vātsyāyana's frame of analogical reasoning and Diṅnaga's frame of the same, is that the relation between signifying analogy and signified analogy in the observed world, as understood by Vātsyāyana, was that of equivalence, while Diṅnāga considers this relation in the world of *sapakṣa* and *vipakṣa*, as that of implication.[17a] The term *avinābhāva* which is supposed to be used by Diṅnāga is suggestive of the same relation which can be formulated "It is not the case that *hetu* can exist without *sādhya*". It may be noticed, however, that in Diṅnāga's frame of *anumāna* this relation between *hetu* and *sādhya* is observed to be real in the world of *sapakṣa* and *vipakṣa*. Thus *pakṣa* is excluded from the scope of this relation. Secondly it seems that *sapakṣa* and *vipakṣa* are the two mutually exclusive and collectively exhaustive sections of the observed world. Thus the process of Diṅnāga's *anumāna* remains to be a strict analogical process from observed to unobserved.

Some evidences may be cited for justifying the same. In *Nyāyapraveśa* by the Diṅnāga's disciple Śaṅkarasvāmin,[18] the word *sapakṣānugama* is used for *anvaya* which restricts the scope of *vyāpti* and keeps the *pakṣa*-case out the the realm of *vyāpti*[19]. This is also frequently seen that Diṅnāga is particular about using the word *dṛṣṭa* (= seen or observed) while giving the examples of *anvaya* or *vyatireka*. TUCCI's translation of *Nyāyamukha* gives many evidences of this kind. See for instance,

'We see that whatever is immaterial is eternal.....',

'We see that whatever is non-eternal is material....'[20]

Here 'we see that' seems to be the translation of *dṛṣṭam*' of Sanskrit original which makes reference to the observed world.

It is also an interesting fact that Diṅnāga did recognise that even though all the three CEs are possessed by a certain *hetu*, the *hetu* can still be inconclusive *(anaikāntika)*. We shall see in 3.24 that the fallacy called *viruddhāvyabhicāri* mentioned by Diṅnāga, occurs when the *hetu* has three CEs but when a contrary thesis can be held with the help of another *hetu* which equally possesses the three CEs. Diṅnāga was unable to show how his frame of sound reasoning was capable of keeping this fallacy away.

Part II

§ 3.12

According to Diṅnāga, *parārthānumāna*, which he also calls *sādhana*, consists of three elements namely *pakṣa* (= thesis), *hetu* (= signifying analogy) and *dṛṣṭānta* (= instance).[21] Out of these elements if any element is defective then the total inference is defective. Therefore Diṅnāga regarded *pakṣābhāsa* (fallacious thesis), *hetvābhāsa* (fallacious *hetu*) and *dṛṣṭāntābhāsa* (fallacious instance) as the types of *sādhanābhāsa* (fallacious proof).[22]

For explaining *pakṣābhāsa, hetvābhāsa* and *dṛṣṭāntābhāsa* we consider Diṅnaga's general form of *parārthānumāna* namely P_1. We do not consider P_2 here because it does not contain the statement of negative instance,[23] and thus will not be useful for explaining the fallacious negative instance. The drawback of P_1 however is that we cannot explain with it the plausibility of the inference from purely positive analogy. P_1 is still useful because we can explain with it all the *sādhanābhāsas* that Diṅnāga refers to.

Out of the three elements of proof, namely *pakṣa, hetu* and *dṛṣṭānta*, the *pakṣa*-element is expressed by the thesis in P_1. The first step in P_1 expresses a part of *hetu*-element. The second and the third steps express the *dṛṣṭānta* as well as the remaining part of the *hetu*-element. A *Pakṣābhāsa* thus affects the truth of the theis. A *Hetvābhāsa* affects the truth of the first or the second or the third step. A *Dṛṣṭāntābhāsa* affects the truth of the second or the third step.

Pakṣābhāsa or a pseudo-thesis is a thesis proposed by an arguer and which is falsified by some authentic cognition or by some universally accepted convention or by itself. Diṅnāga in *Nyāyamukha* gives five types of *pakṣābhāsas* namely *svavacanaviruddha* (falsified by itself), *āgamaviruddha* (contradicted by some other thesis accepted by oneself), *prasiddhiviruddha* (contradicted by some commonly accepted linguistic synonymity[22a]), *pratyakṣaviruddha* (contradicted by a true observation- statement) and *anumānaviruddha* (contradicted by the conclusion of some sound inference).

§ 3.14

Svavacanaviruddha may mean 'opposed by itself' or 'opposed by its utterance.'[23] In the first case the thesis would be self-falsifying, while in the second case it would be a pragmatic contradiction.

We find two types of examples of this fallacy. One is, 'All words are false.'[24] This must be in fact 'All sentences are false.' If the thesis namely "All sentences are false" is true then the thesis being a sentence must be false. In this way the thesis is self-falsifying.

Udyotakara quotes a different example which states 'The words are meaningless.'[25] This thesis is not 'contradicted by itself' in the same way as is the first example. The thesis does not present a logical opposition but a pragmatic contradiction, because the person who tries to communicate something with the help of the sentence namely

'The words are meaningless' assumes that the words which he is utterring at the same time, do mean something.

It is possible, however, that TUCCI's translation of the example is misleading. If the example quoted by Udyotakara is the same as the example originally given by Diṅnāga, then TUCCI's translation must be inauthentic.

§ 3.15

The paseudo-thesis called *āgamaviruddha* is one which is contradicted by some thesis or doctrine already accepted by the arguer. The example of this is the thesis namely 'Sound is eternal' when it is proposed by a Vaiśeṣika.

One who is presenting such a 'pseudo-thesis' may not be presenting a false thesis but he is said to commit the fallacy because he is inconsistent with himself while presenting the thesis. This may be compared therefore with the fault of statement called *siddhāntaviruddha*[26] as well as with the external fallacy of *viruddha.*[27]

§ 3.16

The pseudo-thesis called *prasiddhiviruddha* is one which is contradicted by some commonly accepted linguistic synonymy. The example runs '*Śaśī* is not Candra.' (*Śaśī* and Candra are the synonyms meaning 'moon') In English the example may be stated as 'Luna is not the moon.'

This type of statement may be thought as a weaker kind of the pragmatic contradiction, because while presenting the thesis like 'Luna is not the moon' the arguer is contradicting the commonly accepted conventions of Egnlish language, which he has implicitly accepted while talking in English.

The general notion of *prasiddhiviruddha* as accepted by Diṅnāga, however, is not restricted to the weaker pragmatic contradiction, but it should be interpreted as contradicting with any commonly accepted convention. This notion of the fallacy assumes the public or universal acceptance of a particular convention to be the norm of the acceptability of that convention. Here acceptance is confused with acceptability.

Śaṅkarasvāmin names this fallacy as *loka-viruddha*.[28] His example of this fallacy is worth-noting in this context. The examaple runs.

'The human skull is pious,

because it is a limb of an animal,

like a conch or a shell.'

Such an inference is made by opposing the general opinion that the human skull is not pious. The difficulty in this example is that the belief that the human skull is not pious, is not a universally accepted belief because for instance, the devotees belonging to *śākta* cult are said to believe that the human skull is pious. The belief is not universally accepted though accepted by majority. Acceptance by majority cannot be regarded as a standard of truth. Even the universal acceptance cannot be regarded so.

§ 3.17

The pseudo-thesis called *pratyakṣaviruddha* is falsified by our experience. That is to say, it is contradicted by some true observation-statement.

Dinnāga's example runs, 'Sound cannot be heard.' Our experience shows that sound can be heard. Thus the thesis namely sound cannot be heard, is falsified by our experience.

Udyotakara raises an objection regarding the example.[29] He argues that 'Sound can not be heard' is not falsified by perception but by inference. The objection seems to be sound because a particular sound is perceived but the general fact that 'sound is perceptible' is not perceived, but inferred. In this way 'Sound can be heard' is not an observation- statement.

§ 3.18

The pseudo-thesis called *anumānavirddha* is contradicted by the conclusion of some sound inference. Diṅnāga's example runs, 'The pot is eternal.' Here, a sound inference which proves that the pot is non-eternal, is available to us. This fallacy is distinct from *viruddhāvyabhicāri* because here the inference with opposing conclusion is stronger than the proposed inference. But in *viruddhāvyabhicāri*, both inferences are equally strong and neither inference 'falsifies' the conclusion of the other inference. In case of *anumānaviruddha*, however, the pseudo-thesis is falsified by some established sound inference.

§ 3.19

Śankarasvāmin adds four types of *pakṣābhāsa* to Diṅnāga's list. The additional four types are *aprasiddhaviśeṣaṇa* (unrecognised predicate of the thesis), *aprasiddhaviśeṣya* (unrecognised subject of the thesis), *aprasiddhobhaya* (unrecognised subject and predicate of the thesis) and *prasiddhasambandha* (already accepted predication).[30] One of the first three fallacies occurs when some part of the disputant's thesis is not recognised by the opponent. In that case the disputant has first to prove that part independently of the thesis and then proceed with his thesis. For instance a Buddhist while arguing with a *Sāṅkhya* thinker, may present his thesis that sound gets destroyed. *Sāṅkhya* can claim that this thesis commits the fallacy called *aprasiddhaviśeṣaṇa*, because the predicate of the thesis, namely destruction is not recognised by Sāṅkhya. Sāṅkhya does not accept anything getting destroyed.[30a] Thus the Buddhist has first to demonstrate how the destruction is possible. Then he can proceed to prove that sound gets destroyed. Fourth is the fallacy of superfluous thesis.[31] If the opponent has already accepted the relation between subject and predicate expressed in the thesis, then the disputant's arguing for such a thesis is superfluous.

It may be seen that all the above fallacies of thesis added by Saṅkarasvāmin are significant only in the context of debate. These may be called in this sense 'contextual fallacies.'[32]

§ 3.20

Diṅnāga's account of *pakṣābhāsa* clearly suggests that fallacy of thesis occurs due to falsity of the thesis. It may also be seen that in Diṅnāga's account, the logical consideration is sometimes mixed with the extralogical consideration. Thus his *prasiddhiviruddha* does not exhibit purely logical considerations but exhibits also the extralogical consideration, namely his belief in the commonly accepted conventions. In spite of this, Diṅnaga's notion of *pakṣābhāsa* is inseparable from the notion of the falsity of the thesis. However, Śaṅkarasvāmin's notion of *pakṣābhāsa* deviates from the original notion, because his 'fallacy of superfluous thesis' is a fallacy in spite of the truth of the thesis concerned.

§ 3.21

Falsehood and the possibility of falsehood (i.e. dubitability of truth) are the two basic notions generally concerned with any *sādhanābhāsa*.[33] Out of these basic notions *pakṣābhāsa* is mainly concerned with falsehood and not with possibility of falsehood. Because a thesis is argued for, when its truth is already dubitable. So at the primary stage, the dubitability of the (truth of the) thesis cannot amount to the fallacy of the thesis.[34] But the dubitability of any premise of the argument does amount to the fallacy in that premise, because the premises of any sound argument, which are used to prove the certainty of some thesis, must not themselves be uncertain. Therefore when we are concerned with the fallacies in the premises of the argument, we are concerned with falsehood as well as with dubitability of the premises.

§ 3.22

Out of the three elements of proof, namely *pakṣa*, *hetu* and *dṛṣṭānta*, *pakṣa*-element is expressed by the thesis in P_1 Out of the other two elements, though the first step of P_1 is supposed to express the *hetu*-element, and the remaining two steps namely the second and the third one are supposed to express the *dṛṣṭānta*-element, it is rather difficult to divide the two elements in this manner. For the first step expresses only the first CE of *hetu*, while the second CE and third CE of *hetu* are expressed by the second step and the third step respectively. Thus *hetu*-element is expressed partly by the first step and partly by the second and the third step. The first part of the second step which runs as

If anything (other than *pakṣa*-subject) has *hetu*, then it has *sādhya*

expresses the main part of the second CE namely B_2, while the second part of the second step which is of the form

'For example d (which is other than *pakṣa*-subject) is such that both *hetu* and *sādhya* are predicated of it."

expresses not only a part of second CE, namely B_1, but also the positive instance, which is a part of *dṛṣṭātna*-element. Thus the second CE itself may be divided into *hetu*-element and *dṛṣṭānta*-element. Similarly the third CE itself may be divided into *hetu*-element and *dṛṣṭānta*-element.

If seen from another perspective, the second and the third step in P_1 can be treated as mainly concerned with *dṛṣṭānta*-element (in a wider sense). According to Diṅnāga's frame of *anumāna*, the known world is divided in two parts, *pakṣa* and *dṛṣṭānta*. *Dṛṣṭānta* is the world of observed instances, while *pakṣa* is not fully observed, in the sense in which the world of instances *(dṛṣṭānta)* is observed. It may be seen again that the thesis and the first step are subjected to the *pakṣa*-subject while the second and the third step are subjected to the subjects of instances. *Hetu* is like a mediator which helps us in passing from the world of instances to the *pakṣa*.

Thus in the second step and in the third step *hetu* and *dṛṣṭānta* both are significant. When we discuss the fallacies based on falsity/dubitability of second or third step, sometimes to regard a particular fallacy as *dṛṣṭāntābhāsa* (and not as a *hetvābhāsa)* or as a *hetvābhāsa* (and not as a *dṛṣṭāntābhāsa*) becomes a matter of convenience.

§ 3.23

The fallacies based on falsity or dubitability of the first step, that Diṅnāga refers to, seem to be *svarūpāsiddha* (as it was called later on), *āśrayāsiddha* (or *asiddhadharmī* as termed in *Pramāṇa-Samuccaya*)[35] and *sandigdhāsiddha*.[36] The first step can be false, according to Diṅnāga, either if the subject of the statement, does not exist or if it exists but *hetu* is not predicated of it. We find that *āśrayāsiddha* occurs when the subject does not exist and *svarūpāsiddha* occurs when the subject exists but *hetu* is not predicated of it. It may be noticed here that though the term *svarūpāsiddha* is not used by Diṅnāga to denote any *hetvābhāsa*, his example of *ubhayāsiddha* namely, 'Sound is eternal, because it can be perceived by eyes,' tallies with the popular example of *svarūpāsiddha*. The only distinction seems to be that in *ubhayāsiddha*, the context of debate is taken into consideration, while in svarūpāsiddha, it is not. *Sandigdhāsiddha* is evidently based on the dubitability of the first step.

Diṅnāga's example of *āśrayāsiddha* namely,

'(Somebody discussing with the Buddhist, argues that) the ātman is all-pervading because it produces the feelings like joy'[37]

Suggests that *āśrayāsiddha* refers to 'unproved subject of the thesis' and not to 'undoubtedly absent subject of the thesis.' In this sense the example corresponds to the fallacy based on dubitability and not on falsity. Diṅnāga's example may be thus distinguished from the popular Navya- Nyāya example of *āśrayāsiddha*, namely,

'The sky-lotus is fragrant

because it is a lotus,'

which is based on the non-existence of the subject and thus on the falsity of the first step.

However, this gives a hint that we can classify *svarūpāsiddha* and *āśrayāsiddha* into two subsclasses namely *asiddha* (in the sense 'disproved' or 'false') and *sandigdhāsiddha* (in the sense 'unproved' and 'dubitably true'). *Sandigdhāsiddha*, thus, need not be treated as a special variety of *'asiddha'* (= disproved) but it should be treated as a subvariety of *svarūpāsiddha* as well as *āśrayāsiddha.*

§ 3.24

Though Diṅnāga considers only two ways in which the first step can be false, in fact there are atleast three ways,

1) *Hetu* and *pakṣa-* subject both exist but *hetu* is not predicated of the *pakṣa*-subject. In other words, *hetu* is predicated of some subject other than the *pakṣa*-subject and some other predicate is predicated of the *pakṣa*-subject but *hetu* is not predicated of the *pakṣa*-subject. Though this fallacy is generally called *svarūpāsiddha,* I would like to call it *sambandhāsiddha* and distinguish it from *svarūpāsiddha.*

An example of this is the popular example of *svarūpāsiddha,* which is referred to in the previous section. It is clear in that example that sound exists and there are some things which are perceptible by eyes. Thus *hetu* and *pakṣa*-subject both exist but that *hetu* is predicated of the *pakṣa*-subject is false.

2) *Hetu* exists but *pakṣa*-subject does not. Therefore the predication is false. This is a case of *āśrayāsiddha.*

3) *Hetu* does not exist but *pakṣa*-subject does. Therefore the predication is false. This should be in fact called *svarūpasiddha* because here *hetu* is by its nature *(svarūpa)* non-existent. Its example would be as follows,

The mountain must have fire

because it possesses the horn of a hare.

Here the *hetu,* namely horn of a hare, does not exist though the mountain does. This type of *asiddha,* however, is never seen to be recognised by Indian logicians.

Just as we can conceive of at least three varieties (the forth variety may be the combination of second and the third variety), of *asiddha* based on falsity, similarly we can conceive of at least three varieties of *asiddha* based on dubitability. Here again we find that the variety in which the existence of *hetu* is doubtful, but the existence of *pakṣa*-subject is not, is not found to be recognised by any Indian logician. It seems that the occurrence of such a type of fallacy must be very very rare.

§ 3.25

While classifying *asiddha*, Diṅnāga also refers to two other types namely *ubhayāsiddha* and *anyatarāsiddha*.[38] These types are significant only in a debate-context and in this sense we may call these fallacies as contextual fallacies. Here the principle that governs these types is neither falsity nor dubitability, but the principle of acceptance. The fallacy called *ubhayāsiddha* is committed when the first step of the *parārthānumāṇa*, asserted (but not accepted) by some debater, is not accepted by both the debaters, neither by the disputant nor by the opponent. The fallacy called *anyatarāsiddha* is committed when the first step is not accepted by one of the debaters.

Thus the principle behind these fallacies is that in the debate-context any debater can make a good inference only if its premises are accepted by both the debaters. In this manner, according to Diṅnāga we can conceive of the contextual fallacies of second step and third step as well.[39] It may be seen, however, that the contextual fallacies are many times relative fallacies. In a changing context, the argument which is contextually fallacious in one context can be 'contextually sound' in another context.

§ 3.26

The fallacies of *hetu* based on the falsity of the second and/or third step, according to Diṅnāga, seem to be *sādhāraṇa-anaikāntika* and *viruddha*. *Asādhāraṇa-anaikāntika* is, however, based on the falsity of merely the second step.

While discussing the faliacies of *hetu* based on the falsity of the second and the third step, we have to restate the second and the third step in accordance with their import of *hetu*-element. We have seen in § 3.7, for instance, that the proper expression of B_2 (read with B_1) would be

There exists something (other than *pakṣa*-subject) of which *hetu* and *sādhya* are predicated.................................... (I)

while Diṅnāga chooses to state an instanciation of (I) in place of (I). The instanciation will be of the form -

d (which is other than *pakṣa*- subject) is such that both *hetu* and *sāddhya* are predicated of it................(II)

It is clear that (II) implies (I).

Thus for the sake of our convenience, we admit (I) as the expression of B_2 when we discuss the fallacies of *hetu* based on the falsity of the second step; but we admit (II) as the proper expression of B_2 when we discuss the fallacies of instance based on the falsity of the second step. Thus while considering the fallacies of *hetu* based on the falsity of the second step, our second step is

"If anything (other than *pakṣa* subject) has *hetu* then it has *sādhya.* And there is something (other than *pakṣa*-subject) which has both *hetu* and *sādhya.*"

Similarly while discussing the fallacies of *hetu* based on the falsity of the third step, our third step is

"If anything (other than *pakṣa*-subject) does not have *sādhya* then it does not have *hetu. And there is something (other than pakṣa*-subject) which has neither *sādhya* nor *hetu*"

§ 3.27

By *sādhāraṇa-anaikāntika*, Diṅnāga means a *hetu* which exists both in *sapakṣa* and in *vipakṣa.* If *hetu* exists in *sapakṣa*, then this means that *hetu* partially possesses the second CE, because

"There is something (other than *pakṣa*-subject) which has both *hetu* and *sādhya*" is true. But if *hetu* resides[39a] in *vipakṣa* then this means, that there is something (other than *pakṣa*- subject) which has *hetu* but not *sādhya*, is true. And this is the negation of the first part of the second step of P_1 orP_2.[39b]

Thus, if *hetu* resides in *vipakṣa* then the first part of the second step which is a conjuction of two statements must be false. If any one of the conjuncts is false, then the conjunction is false. Thus if *hetu* resides in *vipakṣa* then the second step must be false. Thus *sādhāraṇa-anaikāntika* implies the falsity of the second step.

Similarly that there is something (other than *pakṣa*- subject) which has *hetu* but not *sādhya* is also the negation of the first conjunct of the third step.[39b]

Thus *sādhāraṇa-anaikāntika* implies the falsity of the third step as well.

§ 3.28

Śaṁkarasvāmin adds three types to the list of *anaikāntika-hetvābhāsas*.[40] The additions are *sapakṣaikadeśavṛtti-vipakṣa-vyāpī (hetu* residing in a part of *sapakṣa* and in the whole *vipakṣa) vipakṣaikadeśavṛtti-sapakṣavyāpī (hetu* residing in a part of *vipakṣa* and in the whole *sapakṣa*) and *ubhayapakṣaikadeśavṛtti (hetu* residing in a part of *sapakṣa* and in a part of *vipakṣa*). Though these 'additions' are not referred to by Diṅnāga as the fallacies distinct from *sādhāraṇa-anaikāntika*, they are considered by him in his *Hetucakra*. It seems that Diṅnāga included all these in *sādhāraṇa- anaikāntika*. Śaṅkarasvāmin meant by *sādhāraṇa-anaikāntika*, only that *hetu* which resides in whole *sapakṣa* and whole *vipakṣa*; which was only a subtype of Diṅnāga's *sādhāraṇa-anaikāntika*. These four types of *sādhāraṇa-anaikāntika* correspond to the four ways in which *hetu* can reside in both *sapakṣa* and *vipakṣa*.

§ 3.29

By *viruddha*, Diṅnāga means that *hetu* which does not reside in any part of *sapakṣa* but resides in *vipakṣa* (partially or wholly).

'*Hetu* does not reside in any part of *sapakṣa*' means,

"If anything (other than *pakṣa*-subject) has *hetu* then it does not have *sādhya*"

And '*hetu* resides in *vipakṣa*' means

"There is something (other than *pakṣa*-subject) which has *hetu* but not *sādhya*"

We have seen in §3.27 that if *hetu* resides in *vipakṣa* then the second step must be false. Similarly, we saw there, that if *hetu* resides in *vipakṣa* then the third step must be false as well. Thus in the similar fashion as *sādhāraṇa-anaikāntika* does, *viruddha* also implies the falsity of the second as well as the third step.

The complete form of *viruddha* is "If anything (other than *pakṣa*-subject) has *hetu* then it lacks *sādhya*. And there is something (other than *pakṣa*-subject) which has *hetu* but not *sādhya*"

This statement can be derived by substituting absence of *sādhya* for *sādhya* in the second step of P_1. Thus the form of *viruddha* indicates that this type of fallacious *hetu* can be in fact used for proving absence of *sādhya* when it is being wrongly used to prove *sādhya*. That is why, this *hetu* is called *viruddha* (= opposite).

Just as we have four types of *sādhāraṇa anaikāntika* according to the alternative situation of *hetu* in *sapakṣa* and *vipakṣa*, similarly we have two types of *viruddha* according to the alternative situation of *hetu* in *vipakṣa*. The *viruddha*-kind of fallacious *hetu* exists in *vipakṣa*. It may exist in a part of *vipaṣa* or in the whole *vipakṣa*. Accordingly the two types of *viruddha* may be named as *vipakṣaikadeśavṛtti* and *vipakṣavyāpī* respectively. Both these types are referred to by Diṅnāga in his *Hetucakra*.

As compared with the types of *viruddha* discussed in *Nyāyamukha*, three types are seen to be added in *Nyāyapraveśa*. The additions are

(i) *dharmaviśeṣaviparītasādhana (hetu* that proves the negation of some implicit characteristic of *pakṣa* - predicate). (ii) *dharmisvarūpaviparītasādhana* (*hetu* that proves the negation of the *pakṣa*-subject) and (iii) *dharmiviśeṣaviparītasādhana (hetu* that proves the negation of some implicit characteristics of the *pakṣa*-subject). Out of the three, first and the third one is supposed to prove the negation of the thesis intended, but not verbally expressed by the debater. In a particular debate-context, a debater A may argue for his thesis X, while he may be finally aiming at proving the thesis X́. The other debater B also may be aware of the aimed thesis of A. Thus B may show that the *hetu* that A is providing may prove X, but at the same time it proves the negation of A's aimed thesis namely X́. These two fallacies are thus, in a peculiar way, significant only in a debate- context. However, these should not be confused with the external fallacy of *viruddha*.[41]

The second addition runs, '*hetu* that proves the negation of *pakṣa*-subject.' We saw (§3.29) that in Diṅnāga's *viruddha*, *hetu* is supposed to prove that *sādhya* is not predicated of *pakṣa*-subject. But that can be proved even if it is shown that *pakṣa*-subject does not exist. Thus if *hetu* proves non-existence of *pakṣa*-subject it automatically proves the negation of the thesis as well. Thus, if we follow Diṅnāga's general description of *viruddha*, this type (namely the second addition) can be treated as a subtype of Diṅnāga's *viruddha*.

§ 3.31

By *asādhāraṇa-anaikāntika*, Diṅnāga means a *hetu* which resides neither in *sapakṣa* nor in *vipakṣa*. If *hetu* does not reside in *vipakṣa*, then it possesses the third CE. But if *hetu* does not reside in any part of *sapakṣa* then it fails to possess the second CE because then,

"There is something (other than *pakṣa*-subject) which has both *hetu* and *sādhya*"

is false, which is our second conjunct of the second step. Thus if *hetu* does not reside in *sapakṣa* then the second step must be false. *Asādhāraṇa-anaikāntika* thus implies the falsity of the second step.

§ 3.32

As we saw in §3.26, *sādhārana-anaikāntika* and *viruddha seem to fall in one group of fallacies and asādhāraṇa-anaikāntika* in another. This classification is based on the particular step or steps that are falsified by the particular fallacy. However, Diṅnāga classifies *sādhāraṇa-anaikāntika, asādhāraṇa-anaikāntika* and *viruddhāvyabhicāri* in one group, namely *anaikāntika*, and classifies *viruddha* in another. This classification is based on the ultimate function of the concerned fallacy. The fallacies that create doubt regarding the truth of the conclusion are classified under *anaikāntika* ('inconclusive'), while *viruddha* forms another group of fallacies because it does not create doubt but creates certainty about the falsity of the conclusion. It should be noticed here that this rule of classification is applied by Diṅnāga only to the *hetvābhāsas* other than *asiddha*.

§ 3.33

We find that Diṅnāga while presenting his list of fallacies, has not considered all the possibilities of third and fourth step being false. For instance, second step is false if *sapakṣa* does not exist. That is to say, if

"There is something (other than *pakṣa*-subject) which has *sādhya*" is false, then

"There is something (other than *pakṣa*-subject) which has both *hetu* and *sādhya*" also is false and thus the second step is false. Similarly if *vipakṣa* does not exist, then,

"There is something (other than *pakṣa*-subject) which has neither *sādhya* nor *hetu*" is false and thus the third step is false. However, Diṅnāga, who considers the fallacy called *āśrayāsiddha* where *pakṣa*-subject does not exist, does not consider those fallacies where *sapakṣa* subjects or *vipakṣa*-subjects do not exist. Consequently he does not consider the fallacy where neither *sapakṣa*-subjects nor *vipakṣa*-subjects exist. This fallacy was later on called as *anupasaṁhārī* by Udyotakara.

We see that Diṅnāga while exposing the general nature of the fallacies of second and third step, states that the fallacies of second and third step can be based on their falsity, dubitability or non-acceptance.[42] However, while giving the details of the fallacies of second and third step, he does not mention the fallacies based on dubitability or non-acceptence. It will be found (§5.18-19) that Dharmakīrti later on pays thorough attention to the fallacies of second and third step based on dubitability.

§ 3.34

Thus we can easily see that generally all *hetvābhāsas*, according to Diṅnāga, are based upon the falsity or dubitability or non-acceptance of some premise of the proof. *Viruddhāvyabhicārī* which Diṅnāga includes in *anaikāntika*, is an exception. *Viruddhāvyabhicārī* means a strong signifying analogy which proves the opposite thesis. Thus the fallacy consists of two *hetus*, which in spite of being equally strong analogies, 'prove' mutually contradictory conclusions. Let us see, for instance, the example quoted in *Nyāyapraveśa*.

(i) Sound is non-eternal, because it is a product, like a pot.

(ii) Sound is eternal, because it is knowable by ears like soundness.

Though Diṅnāga includes both these arguments in the fallacy and adds that both the *hetus* are (collectively) *anaikāntika*, [43] it seems that any debater who commits this fallacy, makes only the first argument, while the other debator who points out this fallacy, makes the second argument. It can be seen, however, from the above example that there is no logical opposition as far as the premises of the first argument and the premises of the second argument are concerned. Though the premises of the two arguments are compatible with each other, their conclusions are contradictory with each other. Secondly it should be noticed that in case of this fallacy, both the *hetus* are asserted to possess all the three CEs.

In the above example, for instance, the first *hetu* namely 'being a product' is predicated of the *pakṣa*-subject namely 'sound.' It is also predicated of the *sapakṣa-* subjects like pot etc. It is not predicated of the *vipakṣa*-subjects like space, atoms etc. The second *hetu* namely 'knowability by ears' is predicated of the *pakṣa*-subject namely sound. It is also predicated of the *sapakṣa*-subject namely 'soundness.' It is not predicated of any *vipakṣa*-subjects like pot etc.

Though both the *hetus* thus possess all the three CEs, we are prevented from concluding that both the arguments must be sound, because the conclusions of the two arguments are contradictory with each other.

The above discussion shows that Diṅnāga must have been aware of the fact that the account of three CEs of *hetu* which he gave, was not perfect so as to guarantee the truth of the conclusion. So Diṅnāga had to go out of the way and declare a new fallacy of *hetu* namely *viruddhāvyabhicārī*. Diṅnāga could not notice the exact drawback of the frame of inference he had accepted. Dharmakīrti could observe the drawback and tried to remove it as well.[44]

§ 3.35

We have seen in §3.22 that the *hetu*-element and *dṛṣṭānta*-element of *parārthānumāna* cannot be completely separated from each other. Therefore while discussing the fallacies of instance, we have to bear in mind that according to Diṅnāga, the *dṛṣṭānta*-element is expressed by the complete second step and the third step of P_1 taken together. Thus we find that some fallacies of instance are based on the falsity of the instanciation-part of the second or third step, while other fallacies of instance are based on the faulty presentation or non-presentation of the implication-part of the second or third step.

§ 3.36

We have seen in §3.8 that the case of positive instance is described as

"d (which is obviously other than *pakṣa*-subject) has both *hetu* and *sādhya*"

Where d stands for the subject of the positive instance.

This is the instanciation-part of the second step of P_1.

This instanciation-part can be false in various ways. However, Diṅnāga refers to three fallacies based on the falsity of this. These fallacies are called *sādhanadharmāsiddha* (The pseudo-positive-instance, in which *hetu*-predicate is not predicated of subject of the instance), *sādhyadharmāsiddha* (The pseudo- positive-instance in which *sādhya*-predicate is not predicated of the subject of the instance) and *ubhayāsiddha* (The pseudo-positive-instance in which neither *hetu*- predicate nor *sādhya*-predicate is predicated of the subject of the instance). It is thus obvious that the fallacy called *sādhanadharmāsiddha* occurs when d does not have *hetu*, *sādhyadharmāsiddha* occurs when d does not have *sādhya*; and *ubhayāsiddha* occurs when d has neither *hetu* nor *sādhya*.

§ 3.37

We have seen in §3.8 that the case of negative instance in the third step of P_1, is described as

"$\bar{d}$ (which is obviously other than *pakṣa*-subject) has neither *sādhya* nor *hetu*"

Where $\bar{d}$ stands for the subject of the negative instance.

This is the instanciation-part of the third step of P_1. This instanciation-part can be false in various ways. Diṅnāga refers to three fallacies based on the falsity of this. These fallacies are called *sādhanāvyāvṛtta* (the pseudo- negative-instance in which *hetu*-predicate is predicated of the subject of the instance), *sādhyāvyāvṛtta* (the pseudo-negative-instance in which *sādhya*-predicate is predicated of the subject of the instance) and *ubhayāvyāvṛtta* (the pseudo-negative-instance in which both *hetu*-predicate and *sādhya*-predicate are predicated of the subject of the instance). It is thus obvious that the fallacy called *sādhanāvyāvṛtta* occurs when $\bar{d}$ does have *hetu*, *sādhyāvyāvṛtta* occurs when $\bar{d}$ does have *sādhya* and *ubhayāvyāvṛtta* occurs when $\bar{d}$ does have both *hetu* and *sādhya*. Here the general assumption (except in case of P2) is that the subject of the instance should exist.[45]

§ 3.38

It is clear that Diṅnāga has not mentioned all the types of the fallacies of instance based on the falsity of instanciation-part. For instance, the instanciation-part in the second step is false if *dṛṣṭānta*-subject is the same as *pakṣa*-subject. We can consider an argument like 'Sound is eternal, because it is audible, like the sound' where the subject of instance is the same as the *pakṣa*-subject. Similar example of pseudo-negative-instance can also be formed.

Diṅnāga's general guide-line that the fallacies of premises can be based either on falsity or dubitability or non-acceptance of some premise, suggests that the fallacies of instance can be based upon the falsity or dubitability or non-acceptance of the instanciation-part of the premises. However, he does not seem to have referred to the fallacies of instance based on dubitability or non-acceptance of the instanciation-part. It will be seen again (in §5.27) that Dharmakīrti referred to the fallacies of instance based on dubitability.

§ 3.39

Now we come to Diṅnāga's those 'fallacies of instance' which cannot be clearly distinguished from the 'fallacies of hetu' as such, for these 'fallacies of instance' are based on non-presentation or mispresentation of nothing but *hetu-sādhya* relation in general.

According to Diṅnāga the *hetu-sādhya*-relation which is observed in the world of instances, [46] should be expressed in an argument, and again, expressed in a fixed manner. In the second step, for instance, the form of expressing the relation should be 'Whatever is seen to have *hetu* is seen to have *sādhya*.' If such expression is omitted and only the instanciation-part is stated, then it would, according to Diṅnāga, amount to the fallacy of instance called *ananvaya* (= the fallacy where the observed positive relation is not expressed). If, on the other hand, the observed positive *hetu-sādhya*-relation is expressed conversely then also there would supposedly be a fallacy of instance. This fallacy was called *viparītānvaya* (= the fallacy where the converse of the observed *hetu-sādhya*-relation is expressed).

Thus in the case of ananvaya, the faulty second step is stated as

"d (which is other than *pakṣa*-subject) has both hetu and sādhya" which is only the second part of the second step in P_1 as well as P_2.

While in the case of *viparītānvaya*, the faulty second step is stated as,

"If anything (other than *pakṣa*-subject) has *sādhya*, then it has *hetu*. For example d (which is obviously other than *pakṣa*-subject) has both *sādhya* and *hetu*"

In an identical way, Dinnāga conceives of the two fallacies of negative instance of which one is based on non-presentation and the other on converse presentation of the negative *hetu-sādhya*-relation. The fallacies were called *avyatireka* (= fallacy where the observed negative *hetu-sādhya*-relation is not expressed) and *viparītavyatireka* (= the fallacy where the converse of the observed negative *hetu-sādhya*-relation is expressed) respectively. Thus in case of *avyatireka*, the faulty third step is expressed as

"d̄ (which is obviously other than *pakṣa*-subject) has neither *sādhya* nor *hetu*" which is only the second part of the third step of P_1 while in case of viparītavyatireka, the faulty third step is expressed as

"If anything (other than *pakṣa*-subject) does not have *hetu*, then it does not have *sādhya*. For example d̄ (which is obviously other than *pakṣa*-subject) has neither *sādhya* nor *hetu*"

§ 3.40

Though the argument with any such faulty second or third step, as spelled out in § 3.39, would necessarily be an invalid argument, it would be improper to call the fallacies of instance, referred to there, as 'formal fallacies.' We have already seen[47] that Dinnāga's frame of inference is that of analogical reasoning and logically speaking, Diṅnāga's form of *parārthānumāna* (P_1 or P_2) is an invalid form. Therefore there is no propriety in talking of 'formal fallacies spelled out by Diṅnāga' when Diṅnāga's frame of 'good' inference itself is a frame of invalid inference. However, when we are talking of fallacies in such a context we are talking of the fallacies of analogical reasoning and there we can talk of 'syntactical fallacies' of analogical reasoning if not of 'formal fallacies' in the strict sense assumed in formal logic. A syntactical fallacy of analogical reasoning is one which breaks the necessary syntactical conditions of a strong analogical reasoning.

In this context we can compare Diṅnāga's syntactical structure of analogical argument with the earlier structures. We find that Caraka's analogical reasoning (as stated in §1.5) is potentially of the form of the argument with *ananvaya* and *avyatireka*, as no general statement refering to the observed world of instances occurs in his five-membered inferencial statement. We have also seen (in §2.8) that according to Vātsyāyana, even the converse of *anvaya*- statement or *vyatireka*-statement may be stated in the course of the argument. That is why, Vātsyāyana expects observed equivalence between *hetu* and *sādhya*. Vātsyāyana's form of inference thus resembles the forms of inference with *viparītānvaya* and *viparītavyatireka*. Vātsyāyana was not completely mistaken because along with the statement of the converse of *anvaya* or *vyatireka* he also expected the statement of *anvaya* or *vyatireka* as a part of his five- membered inferencial statement.

Thus the fallacies of instance namely *ananvaya*, *viparītānvaya*, *avyatireka* and *viparītavyatireka*, as spelled out by Diṅnāga, are significant in the context of the forms of analogical reasoning accepted by earlier logicians.

Notes

1. According to Diṅnāga's interpretation of the Vaiśeṣikas' account of *hetvābhāsa*, Vaiśeṣikas classify *hetvābhāsa* into three classes namely *aprasiddha*, *asan* and *sandigdha*. While criticising Vaiśeṣika-account, Diṅnāga roughly equates *aprasiddha* with

asiddha, asan with *viruddha* and *sandigdha* with *anaikāntika*. (See *VK* p. 199). It may be guessed therefrom that Buddhist logicians like Vasubandhu and Diṅnāga may have been inspired by the account of *hetvābhāsa* that appears in *VS*.

2. As correctly pointed out by RANDLE, Kaṇāda must have stated and exemplified only two *hetvābhāsas* and not three as the later commentators supposed. (See ILES p. 190 fn. 2 and p.191). Thus Diṅnāga may have misinterpreted Kaṇāda to have described three *hetvābhāsas* in *VS*. 3.1.10 (Also see Note 1)

3. In *Tarkaśāstra* (a Buddhist text of logic, attributed to Vasubandhu), for instance, the three conditions of *hetu* are referred to as *pakṣadharma, sapakṣasattva* and *vipakṣavyāvṛtti*. See *Tarkaśāstra*, p. 14, included in *PBTL*

4. There is a controversy regarding *Nyāyapraveśa*, whether Diṅnāga composed it or his disciple Śaṅkarasvāmin composed it. I have assumed the latter possibility, which has been generally accepted by the scholars today. In any case it can be assumed, as I have done here, that the general nature of *anumāna* described in *Nyāyapraveśa* was in tune with the same appreciated by Diṅnāga.

5. अनुमानं लिङ्गादर्थदर्शनम्, लिङ्गं पुनस्त्रिरूपम् उक्तम् । *NPS* p.7.

6. See Note 18 in Chapter I. Of course, *pakṣa*, according to Diṅnāga must have other qualifications also, namely it should not be falsified by itself or by perception or by inference etc.

6a. The *sapakṣa* and *vipakṣa* can be explained in terms of class. But the term 'class' also means *jāti*, and *jāti* is altogether denied by all the Buddhist logicians. Therefore I have avoided the use of the term class in this context. Also see Appendix II.

7. Later on, probably at the time of Dharmakīrti, the second CE was described as *'sapakṣe eva sattva'* in which *eva* conveyed the meaning of 'only.'

7a. This term will be explained in §3.39.

8. The 'necessity' here is not the logical necessit but the empirical necessity.

9. तस्मात् प्रतिबन्धप्रसाधकप्रमाणवृत्त्यैव भावाभावौ सिध्यतः इति विप्रतिपत्तिनिरासार्थम् अस्माभिः निश्चितग्रहणं कृतम् । *HBT* Appendix 7, p.72

Here Dharmakīrti is refering to the verse, अनुमेयेऽथ तत्तुल्ये सद्भावो नास्तिताऽसति । निश्चिता ...... see *HBT*, p.222

9a. *Niścita* can either mean certainty or necessity. We have to make a distinction between the two meanings. Certainty is more an epistemological concept while necessity is more a logical one. The menaing of *niścita*, however, comes nearer to that of 'certain' while the meaning of *niyata* comes nearer to that of 'necessary.'

10. See *BL* p.320.

11. ननु च सपक्ष एव सत्त्वम् इत्युक्ते विपक्षे असत्त्वम् एव इति गम्यते एव तत्किमर्थं पुनरुभयोरुपादानं कृतम् ? उच्यते । अन्वयो व्यतिरेको वा नियमवानेव प्रयोक्तव्यः नान्यथा इति दर्शयितुं द्वयोरपि उपादानं कृतम् । *NBT* pp. 109-10.

12. हेतोः सपक्षासपक्षयोः भावाभावौ न परस्परम् आक्षिपतः । *HBT*, Appendix 7,p.72.

13. Though the term *kevalānvayi* does not appear in Diṅnāga's work, Diṅnāga refers to an inference which suits to Udyotakara's description of *kevalānvayi* type of inference (i.e. the inference from purely positive analogy). Diṅnāga's example runs, 'Sound is non-eternal like the pot unlike the space etc.' Here according to the Buddhists space does not exist. See *NM* pp. 27, 37. Also see Fragment 12 in 'Sanskrit Fragments of the *Pramāṇasamuccaya*' by KATSURA, Shoryu, published in JIP Vol. 3, Nos. 1/2, pp. 75-76.

14. Same is not the case with B_2. We cannot assert B_2 unless we are in a position to assert the existence of some *sapakṣa*- subject.

15. *NM* p. 27.

16. See *NM* p. 37.

16a. For symbolic representation of P_1 and P_2 see Appendix III.

17. For 'purely positive analogy' see § 4.4

17a. It is true that implication or equivalence is not an ontological relation. But when we are talking of implication/equivalence between *hetu* and *sādhya* or between one analogy and the other, we are taking *hetu, sādhya* etc. to stand for eliptical sentences.

18. See *NPS* p.2.

19. Such a *vyāpti* was later on called *bahirvyāpti* (i.e., the *hetu-sādhya*-relation known outside the *pakṣa*-case) and was contrasted with *antarvyāpti* (i.e., the *hetu-sādhya*-relation existing in *pakṣa*-case).

20. *NM* p. 41. Also see the use of *dṛṣṭam* in *NPS* p. 2.

21. तत्र पक्षादियचनानि साधनम् । पक्षहेतुदृष्टान्तवचनैर्हि प्राश्निकानाम् अप्रतीतो ऽर्थः प्रतिपाद्यत इति । *NPS* p.7.

22. एषां पक्षहेतुदृष्टान्ताभासानां वचनानि साधनाभासम् । *NPS* p.7

22a. The linguistic synonymy is always based upon convention. Thus *prasiddhiviruddha* ultimately means that which is contradicted by some commonly accepted convention.

23. For P_1 and P_2 see § 3.8 and § 3.9 respectively.

24. See *NM* p. 7

25. *'avācakāḥ śabdāḥ,' NV* p. 113.

26. See § 1.7
27. See § 2.14
28. See *NPS* p.2
29. See *NV* p. 113.
30. See *NPS* pp. 2-3.

30a. Of course here destruction means annihilation. Sāṅkhyas are prepared to accept transformation of a thing. Thus they would say that a pot may get 'destroyed' as a pot, that is to say, it may get transformed into a non-pot; but still it cannot get destroyed (annihilated); it still persists as earth.

31. For the 'fallacies of superfluity' see § 8.48.
32. For 'contextual fallacies' see § 8.49.
33. 'Invalidity' also is a basic notion. But it is not significant when we consider Diṅnāga's notion of *sādhanābhāsa*. See §3.40.
34. Sometimes at the secondary stage, that is when the thesis has already been argued for, the dubitability of the thesis can be shown with the help of counter-argument. At this stage we can think of the fallacy of thesis based on dubitability. *Viruddhāvyabhicāri* or *satpratipakṣa* in this sense can be called as a fallacy of thesis.
35. See *NM* p. 14, fn. 26.
36. Ibid, fn. 25.
37. See *NM* p. 14.
38. See *NM* p..14 fns. 23 and 24.
39. Diṅnāga says, 'Concerning the (other characteristics of the reason), that is residence or non-residence of the property to be proved in the positive instances etc., the same rules (as those already stated) must be applied' *NM* p. 15. Earlier Diṅnāga has referred to the rules of the first CE (or the first step) that the first step should not be either disproved or unproved (doubtful) or rejected by any debater. (See *NM* p. 14). For the contextual fallacies see § 8.49.

39a. The Indian logic is inseparable from ontology. Therefore the model of residence or existence of *hetu* in *pakṣa/sapakṣa/vipakṣa*-case, which is an ontological model, is always found used in Indian logical discussions.

39b. For symbolic presentation of second and third step in P_1 and P_2 see Appendix III, Part (2). It can be confirmed in terms of the symbolisation there that *sādhārana-anaikāntika* is equivalent with the negation of B_2 as well as with that of C_2.

40. See *NPS* pp. 3-4.

41. The difference between the two is that in case of the external fallacy of *viruddha*, the thesis contradicted by the *hetu* is previously accepted by the debater, while in case of the above-discussed fallacies, the *hetu* proves the opposite of the 'intended thesis.' For external fallacy of *viruddha*, See § 2.14.

42. See note 39 of this Chapter.

43. उभयोः संशयहेतुत्वाद् द्वावप्येतावेकोऽनैकान्तिकः समुदितावेव । *NPS* p.5. Compare this with सोऽयमुभयपक्षसमो विशेषहेत्वभावः इत्युभयमप्यनैकान्तिकम् । *NBH* on *NS* 5.1.39. Diṅnāga may have been inspired by Vātsyāyana's explanation of *jāti* while describing *viruddhāvyabhicāri*.

44. See § 5.2-5.

45. Praśastapāda takes this assumption seriously when he accepts *āśrayāsiddha* as a kind of pseudo-instance. See *PB*, p. 123. Thus he rules out the acceptability of *kevalānvayi* inference.

46. See § 3.22.

Chapter IV

Udyotakara on Anumāna and Hetvābhāsa

Part I
§ 4.1

When Diṅnāga presented a thorough and systematic account of *anumāna* (and also of *pratyakṣa*), Nyāya logicians after Diṅnāga had to take a serious note of this. As the socio-cultural antagonism was developing amongst the Nyāya and Buddhist schools of thought, it is probably because of such extralogical reasons that Udyotakara in *Nyāyavārtika* took a great interest in criticising Diṅnāga and his followers. It is also seen that Udyotakara does not engage himself so much in the rational justification of Nyāya view as he does in criticising Buddhist view.

As far as the nature of *anumāna* is concerned, Udyotakara raises two types of objections against Diṅnāga. The first type of objections can be called verbal and superficial; but the second one is positively philosophical. In the first part of this chapter we are going to discuss Udyotakara's position concerning the status of *anumāna* and his important philosophical objections against Diṅnāga.

Though it has been stated by many scholars of Indian Logic[1] that the concept of pervasion (*vyāpti*) occurs first in the treatise of Udyotakara, it should be examined whether Udyotakara really thought in terms of universal and invariable *hetu-sādhya*-relation. Udyotakara, following Vātsyāyana, talks of *liṅgaliṅgisambandha*. But again, it is the relation between *hetu* and *sādhya* noticed in the observed world. While describing *liṅgaliṅgisambandhasmṛti* (= the recollection of the *hetu-sādhya*-relation), Udyotakara says, '*yatra dhūmam adrākṣaṁ tatrāgnim adrākṣam ityetasyāḥ ca smṛteḥ* .....'[2]. Thus the form of such recollection is 'wheter I saw *hetu* there I saw *sādhya*', which clearly refers to the observed world, and does not at all refer to the 'whole world'. Though Udyotakara occasionally uses the terms like *avinābhāva* to denote the *hetu-sādhya-* relation, he restricts the scope of *avinābhāva* to the *sapakṣa*-cases. For instance he says,

तादृगविनाभाविधर्मोपदर्शनं हेतुरित्यपरे । तादृशा विना न भवति इत्यनेन किल द्वयं लभ्यते विपक्षासत्त्वं तादृशसत्त्वं च। [3]

Here *tādṛk* or *tādṛśa* means *sapakṣa*.[4] Thus Udyotakara's notion of *liṅgaliṅgisambandha* or *avinābhāva* refers to the *hetu-sādhya*-relation in the observed world. Diṅnāga's basic frame of inference, as strong analogical reasoning, therefore, is not overthrown by Udyotakara. On the contrary, Udyotakara sticks to Diṅnāga's central notion of *anumāna*.

§ 4.3

Udyotakara, however, raises an important philosophical objection against Diṅnāga's formulation of the three CEs of *hetu*. While presenting the CEs of *hetu*, Diṅnāga had assumed that corresponding to each sound *hetu* there is a *pakṣa*-subject, at least one *sapakṣa*-subject, and at least one *vipakṣa*-subject (with some exceptions[5]). Though Udyotakara was sure that, for there being a sound *hetu*, the existence of *pakṣa*-subject is essential, however, he thought that the existence of both the remaining two types of subjects namely *sapakṣa*-subject and *vipakṣa*-subject, is not essential. A *hetu* can be a good *hetu* either without *sapakṣa* or without *vipakṣa*. The good *hetu* which has *pakṣa* and *sapakṣa* but does not have *vipakṣa*, was called *anvayī hetu* by Udyotakara. Similarly the good *hetu* which was only *pakṣa* and *vipakṣa*, but does not have *sapakṣa* was called *vyatirekī hetu* by him.[6] *Anvayī* was later on named as *kevalānvayī* and *vyatirekī* as *kevalavyatirekī*. The good *hetu* which has all the three, namely *pakṣa*, *sapakṣa* and *vipakṣa* was called *anvayavyatirekī* by him.

§ 4.4

As far as *anumāna* is considered as strong analogical reasoning, Udyotakara seems to be right in pointing out the cases, where

Diṅnāga's strict frame of *anumāna* in which all the three CEs are essential, cannot be applied.

We can understand *anvayī hetu* as 'purely positive signifying analogy' and *vyatirekī hetu* as 'purely negative signifying analogy'. In the usual analogical reasoning the same signifying analogy can serve as the positive analogy between *pakṣa* and *sapakṣa* and as the negative analogy between *pakṣa* and *vipakṣa*. For instance when we infer 'the fire on the mountain' from the *hetu* namely 'the smoke on the mountain', the smoke serves as the positive analogy between the mountain and the kitchen, and the same smoke serves as the negative analogy between the mountain and the waterful lake. But *anvayī* or *kevalānvayī hetu* is one which can function only as the positive analogy between *pakṣa* and *sapakṣa*, but cannot function as the negative analogy between *pakṣa* and *vipakṣa*, because corresponding to this analogy there is no *vipakṣa*. When, for instance, a Buddhist infers that sound is impermanent because it is a product, the *hetu* namely 'being a product' is the positive analogy between the sound and all other things. But this *hetu*, when presented by Buddhists, cannot serve as a negative analogy because the *hetu* has no *vipakṣa*, as Buddhists do not believe in anything being permanent.

Similarly *vyatirekī* or *kevalavyatirekī hetu* is one which can serve only as the negative analogy between *pakṣa* and *vipakṣa*, but cannot serve as the positive analogy between *pakṣa* and *sapakṣa* because this *hetu* has no *sapakṣa*. When, for instance, a Naiyāyika infers that any living body has a soul, because it contains *prāṇa* etc. then the *hetu* namely 'containing *prāṇa* etc.,' can serve as a negative analogy between *pakṣa* (= living bodies) and *vipakṣa* (the bodies which have no soul), but cannot serve as a positive analogy between *pakṣa* and *sapakṣa*, because here *sapakṣa* (namely the non- living bodies which have soul) does not exist.

§ 4.6

These two special types of *hetu* namely purely positive and purely negative, were later on accepted by almost all the Naiyāyikas and thus the nature of *anumāna* became more and more complicated especially in the hands of Navya- Naiyāyikas. Thus the introduction of these two as the special types of *hetu* (and also the corresponding special types of inference) was an important positive contribution of Udyotakara to the thought concerning *anumāna*. Though previously Diṅnāga had come across the inferences which are made from purely positive analogy,[7] he did not think that the general frame of inference has to be changed for accommodating such inferences. Diṅnāga also overlooked the possibility of the inference from purely negative analogy.

§ 4.7

These types are important when we consider *anumāna* as analogical reasoning. Analogical reasoning is not a logically valid type of reasoning. And when we consider *anumāna* as the valid reasoning based on universal law, we need not give any special importance to these special types of sound *hetu*. Supposing that we consider as example of a valid inference made with a *kevalanyanī hetu*, the *vyaptī* stated in the inference may be called *kevalānvayi vyāpti*. Though it is generally supposed that *kevalānvayi vyāpti* does not have the corresponding *vyatireki vyāpti*, the statement of *kevalānvayi vyāpti* does imply logically the statement of formal *vyatireki vyāpti*, though that *vyatireki vyāpti* may not have any concrete instanciation. Similarly, in case of *kevalavyatireki hetu*, the statement of *kevalavyatireki vyāpti* does imply logically the statement of formal *anvayi vyāpti*.[7a]

Udyotakara's introduction of *kevalavyatireki hetu* as a special type of *hetu* creates a problem when it is compared with the *hetvābhāsa* namely *asādhāraṇa*. This point will be discussed in § 6.24.

Part II

§ 4.8

Udyotakara had made an important contribution to the classification of the fallacies of *hetu*. Udyotakara thinks that in fact *hetu* can be fallacious in innumerable ways.[8] However we can count some of the possible factors which can make a *hetu* fallacious.

Udyotakara is more keen than Diṅnāga as far as he conceives of the different subtle factors which make a *hetu* fallacious. As has been pointed out (in § 3.33), Diṅnāga does not conceive the possibility of the reasoning in which either *sapakṣa*-subjects or *vipakṣa*-subjects or both do not exist. He takes for granted that in any reasoning, good or fallacious, the *sapakṣa* and *vipakṣa* must be existing (or at least conceivable). Udyotakara, however, considers those cases where *sapakṣa* or *vipakṣa* or both are non- existent. However, while conceiving of different types of fallacious *hetu*, he uses the method similar to one used by Diṅnāga. Udyotakara's principle in accounting for the fallacies of *hetu* seems to be as follows: 'Any *hetu* used in an argument is fallacious if any of the premises of the argument is either false or dubitable or is rejected by some debater or doubted by some debater or is superfluous'. In the context of Udyotakara's concept of inference the premises of an argument (*parārthānumāna*) are supposed to express at least the three assertions about the *hetu*.

1) *Hetu* exists in *pakṣa*.

2) Either *sapakṣa* or *vipakṣa* must exist and if *sapakṣa* exists then *hetu* exists in *sapakṣa*.

3) If *vipakṣa* exists then *hetu* does not exist in *vipakṣa*.

These are the states of *hetu* expressed by the premises of the argument with respect to *pakṣa, sapakṣa* and *vipakṣa* respectively. But the actual states of the *hetu* may be different from the expressed states. In that case, if the expressed state or states are defective (that is to say, if they are false or dubitable etc.) then the corresponding *hetu* is called fallacious.

§ 4.9

The method of enumerating the classes of fallacies seems to be as follows.

We first count the varieties of

(1) the states of *hetus*[9] with respect to *pakṣa*.

(We call these states as A-factors. The varieties of A- factors will be enumerated as A_1, A_2 etc.)

(2) the states of *hetus* with respect to *sapakṣa* (We call these states as B-factors. The varieties of B-factors will be enumerated as B_1, B_2 etc.)

(3) the states of *hetus* with respect to *vipakṣa* (We call these states as C-factors. The varieties of C-factors will be enumerated as C_1, C_2 etc.)

All these factors are the actual states of the expressed *hetus*. While deciding whether a particular factor causes a fallacy or not, we compare it with the corresponding expressed state of *hetu*. Given any *hetu* in any *anumāna*, we get some A-factor, some B-factor, some C-factor with reference to that *hetu*. By combining these factors we get the actual description of that *hetu* with respect of *pakṣa, sapakṣa* and vipakṣa. The combination will be referred to in the form,

$Ax\ By\ Cz$

If we have over all n number of A-factors, m number of B- factors and l number of C-factors, then by the rule of combination, we get over all n x m x l combinations. By this method we can count the total varieties of *hetu*. We have to divide this list into two groups: 1) The combinations that refer to good *hetu*. 2) The combinations that refer to fallacious *hetu*. A combination of the actual states of *hetu* is supposed to refer to a good *hetu*, if the expressed states of the *hetu* are in coherence with it. If on the other hand the expressed states of *hetu* do not tally with actual states, and are defective in the sense stated in § 4.8, then that combination of the actual states is supposed to refer to a fallacious *hetu*.

Udyotakara enumerates these varieties of *hetu* in three stages. We call these stages as first classification, second classification and third classification.

§ 4.10

First Classification

(1) A-factors

A_1 :*Hetu* exists in whole *pakṣa*-subject. (*sādhyavyāpaka*).

A_2 :*Hetu* exists in a part of *pakṣa*-subject.[10] (*sādhyaikadeśavṛttidharma*).

A_3 :*Hetu* exists in no *pakṣa* (*sādhyāvṛttidharma*).

(ii) B-factors

B_1 :*Hetu* exists in the whole *sapakṣa* (*tajjātīyavyāpaka*).

B_2 :*Hetu* exists in a part of *sapakṣa*. (*tajjātīyaikadeśavṛtti*).

B_3 :*Hetu* exists in no *sapakṣa*-case. (*tajjātīyāvṛtti*)

B_4 :*Sapakṣa* is non-existent. (*avidyamānatajjātīya*)

(III) C-factors

C_1 :*Hetu* exists in whole *vipakṣa*. (*vipakṣavyāpaka*).

C_2 :*Hetu* exists in a part of *vipakṣa*. (*vipakṣaikadeśavṛtti*).

C_3 :*Hetu* exists in no *vipakṣa*-subject. (*vipakṣāvṛtti*).

C_4 :*Vipakṣa* is non-existent. (*avidyamānavipakṣa*).

§ 4.11

In the first classification thus, we get 3 x 4 x 4 = 48 combinations of the form Ax By Cz. Out of these, only five combinations stand for good (non-fallacious) forms of *hetu*. These are as follows,

A_1 B_1 C_3, A_1 B_2 C_3, A_1 B_1 C_4, A_1 B_2 C_4, A_1 B_4 C_3.

Out of these, first two refer to positive-negative (*anvayi-vyatireki*) *hetu*, Next two refer to the two types of purely positive *hetu*. And last combination refers to purely negative *hetu*. The other 43 combinations refer to fallacious *hetus*, which include different types of *svarūpāsiddha* (A_3 B_1 C_3 etc.), *viruddha* (A_1 B_3 C_1 and A_1 B_3 C_2), *anaikāntika* (*sādhāraṇa* : A_1 B_1 C_1, A_1 B_1 C_2 etc.; *asādhāraṇa* : A_1 B_3 C_3: *anupasaṁhārī* : A_1 B_4 C_4) and the combinations of these fallacies (for instance, A_3 B_3 C_1 as a combination of *svarūpāsiddha* and *viruddha*).

Out of the factors in the first classification, Diṅnāga had not cosidered A_2, B_4 and C_4. Though Diṅnāga considered A_3 while discussing the fallacy of *hetu* namely *asiddha*, in his *hetucakra*[11] (wheel of *hetus*) he considered only A_1 as the A-factor. Thus Diṅnāga's *hetu- cakra* consisted of the following nine combinations,

A_1 B_1 C_1, A_1 B_2 C_1, A_1 B_3 C_1, A_1 B_1 C_2, A_1 B_1 C_3, A_1 B_2 C_2, A_1 B_2 C_3, A_1 B_3 C_2, A_1 B_3 C_3.

§ 4.12

Before we come to the second stage of classification, we may consider the procedure which Udyotakara has followed while giving second and third classification. In addition to the defective[11a] A-factors in the first classification (namely A_2 and A_3), Udyotakara conceives of different 'elementary defective states of *hetu* with respect to *pakṣa*' and constructs the additional A- factors by combining the A-factors with the 'elementary defective states of *hetu* with respect of *pakṣa*'. Here we symbolise the 'elementary defective states of *hetu* with respect of *pakṣa*' so that it will be convenient to enumerate the combinations of these made by Udyotakara, with the A-factors of the first classification.

$\bar{A}_1$: Subject of *hetu*[12] exists in no *pakṣa*. (*viśeṣyāsiddha*)

$\bar{A}_2$: Predicate of *hetu* exists in no *pakṣa*. (*viśeṣaṇāsiddha*)

$\bar{A}_3$: Subject of *hetu* is superfluous. (*asamarthaviśeṣya*)

$\bar{A}_4$: Predicate of *hetu* is superfluous. (*asamarthaviśeṣaṇa*)

$\bar{A}_5$: Subject of *hetu* is dubitable. (*sandigdhaviśeṣya*)

$\bar{A}_6$: Predicate of *hetu* is dubitable. (*sandigdhaviśeṣaṇa*)

$\bar{A}_7$: Subject of *hetu* exists somewhereelse than in *pakṣa*- subject. (*vyadhikaraṇaviśeṣya*)

$\bar{A}_8$: Predicate of *hetu* exists somewhereelse than in *pakṣa*-subject. (*vyadhikaraṇaviśeṣaṇa*)

$\bar{A}_9$: *Hetu* is not accepted by the arguer himself. (*vādyasiddha*)

$\bar{A}_{10}$: *Hetu* is not accepted by the opponent. (*prativādyasiddha*)

$\bar{A}_{11}$: *Hetu* is doubted by either of the two debaters. (*anyatarasandigdha*)

$\bar{A}_{12}$: *Hetu* is doubted by both the debaters. (*ubhayasandigdha*)

$\bar{A}_{13}$: Subject of *hetu* is not accepted by either of the debaters. (*anyatarāsiddha-viśeṣya*)

$\bar{A}_{14}$: Predicate of *hetu* is not accepted by either of the debaters. (*anyatarāsiddhaviśeṣaṇa*)

$\bar{A}_{15}$: Subject of *hetu* is not accepted by any of the two debaters (*ubhayāsiddhaviśeṣya*)

$\bar{A}_{16}$: Predicate of *hetu* is not accepted by any of the two debaters (*ubhayāsiddhaviśeṣaṇa*)

$\bar{A}_{17}$:The *pakṣa*-subject is not proved to exist. (*āśrayāsiddha*)

$\bar{A}_{18}$: *Hetu* can be explained otherwise. (*anyathāsiddha*)

$\bar{A}_{19}$: Subject of *hetu* proves the negation of *sādhya*. (*viruddhaviśeṣya*)

$\bar{A}_{20}$: Predicate of *hetu* proves the negation of *sādhya*. (*viruddhaviśeṣaṇa*)

The above list of 'elementary defective states of *hetu*' shows that the principles of dubitability, non-acceptance and superfluity are regarded as basic principles of fallacies.

§ 4.13

Second Classification.

(I) A-factors

A_1, A_2, A_3 (as stated in § 4.10)

A_4: $A_1 . \bar{A}_1$; A_5: $A_1 . \bar{A}_2$; A_6: $A_2 . \bar{A}_1$; A_7: $A_2 . \bar{A}_2$; A_8: $A_1 . \bar{A}_3$; A_9: $A_1 . \bar{A}_4$; A_{10}: $A_2 . \bar{A}_3$; A_{11}: $A_2 . \bar{A}_4$

(II) B-factors

B_1, B_2, B_3, B_4 (as stated in § 4.10)

(III) C-factors

C_1, C_2, C_3, C_4 (as stated in § 4.10)

Thus in the second classification we get 11 x 4 x 4 = 176 combinations. Udyotakara says, सामान्यतः साध्यवस्तूपग्रहेण हेतुहेत्वाभासभेदः प्रपञ्च्यमानः षट्सप्ततिशतम् ।[13]

Vācaspatimiśra counts these combinations as 48 (of the first classification) + (64 + 64) (of the second classification) = 176, as he says, तदेतत् चतुष्षष्टिद्वयमष्टाविंशतिकं शतं पूर्वया त्रिषोडश्या सह षट्सप्ततं शतम् ।[14]

§ 4.14

Third Classification

(I) A-factors

A_1 to A_{11} (as accepted in § 4.13)

A_{12}: $A_1 . \bar{A}_6$; A_{13}: $A_2 . \bar{A}_6$; A_{14}: $A_1 . \bar{A}_5$;

A_{15}: $A_2 . \bar{A}_5$; A_{16}: $A_1 . \bar{A}_9$; A_{17}: $A_2 . \bar{A}_9$;

A_{18}: $A_1 . \bar{A}_{10}$; A_{19}: $A_2 . \bar{A}_{10}$; A_{20} : $A_1 . \bar{A}_{12}$;

$A_{21} : A_{2}. \bar{A}_{12}$; $A_{22} : A_{1}. \bar{A}_{11}$; $A_{23} : A_{2}. \bar{A}_{11}$;

$A_{24} : A_{1}. \bar{A}_{8}$; $A_{25} : A_{2}. \bar{A}_{8}$; $A_{26} : A_{1}. \bar{A}_{7}$;

$A_{27} : A_{2}. \bar{A}_{7}$; $A_{28} : A_{12}. \bar{A}_{8}$; $A_{29} : A_{13}. \bar{A}_{8}$;

$A_{30} : A_{14}. \bar{A}_{7}$; $A_{31} : A_{15}. \bar{A}_{7}$; $A_{32} : A_{8}. \bar{A}_{8}$;

$A_{33} : A_{10}. \bar{A}_{8}$; $A_{34} : A_{9}. \bar{A}_{7}$; $A_{35} : A_{11}. \bar{A}_{7}$;

$A_{36} : A_{1}. \bar{A}_{13}$; $A_{37} : A_{1}. \bar{A}_{15}$; $A_{38} : A_{2}. \bar{A}_{13}$;

$A_{39} : A_{2}. \bar{A}_{15}$; $A_{40} : A_{1}. \bar{A}_{14}$; $A_{41} : A_{1}. \bar{A}_{16}$;

$A_{42} : A_{2}. \bar{A}_{14}$; $A_{43} : A_{2}. \bar{A}_{16}$; $A_{44} : A_{24}. \bar{A}_{17}$;

$A_{45} : A_{25}. \bar{A}_{17}$; $A_{46} : A_{26}. \bar{A}_{17}$; $A_{47} : A_{27}. \bar{A}_{17}$;

$A_{48} : A_{12}. \bar{A}_{17}$; $A_{49} : A_{13}. \bar{A}_{17}$; $A_{50} : A_{14}. \bar{A}_{17}$;

$A_{51} : A_{15}. \bar{A}_{17}$; $A_{52} : A_{8}. \bar{A}_{17}$; $A_{53} : A_{9}. \bar{A}_{17}$;

$A_{54} : A_{10}. \bar{A}_{17}$; $A_{55} : A_{11}. \bar{A}_{17}$; $A_{56} : A_{24}. \bar{A}_{18}$;

$A_{57} : A_{25}. \bar{A}_{18}$; $A_{58} : A_{26}. \bar{A}_{18}$; $A_{59} : A_{27}. \bar{A}_{18}$;

$A_{60} : A_{12}. \bar{A}_{18}$; $A_{61} : A_{13}. \bar{A}_{18}$; $A_{62} : A_{14}. \bar{A}_{18}$;

$A_{63} : A_{15}. \bar{A}_{18}$; $A_{64} : A_{8}. \bar{A}_{18}$; $A_{65} : A_{9}. \bar{A}_{18}$;

$A_{66} : A_{10}. \bar{A}_{18}$; $A_{67} : A_{11}. \bar{A}_{18}$; $A_{68} : A_{44}. \bar{A}_{18}$;

$A_{69} : A_{45}. \bar{A}_{18}$; $A_{70} : A_{46}. \bar{A}_{18}$; $A_{71} : A_{47}. \bar{A}_{18}$;

$A_{72} : A_{48}. \bar{A}_{18}$; $A_{73} : A_{49}. \bar{A}_{18}$; $A_{74} : A_{50}. \bar{A}_{18}$;

$A_{75} : A_{51}. \bar{A}_{18}$; $A_{76} : A_{52}. \bar{A}_{18}$; $A_{77} : A_{53}. \bar{A}_{18}$;

$A_{78} : A_{54}. \bar{A}_{18}$; $A_{79} : A_{55}. \bar{A}_{18}$; $A_{80} : A_{24}. \bar{A}_{20}$;

$A_{81} : A_{25}. \bar{A}_{20}$; $A_{82} : A_{26}. \bar{A}_{20}$; $A_{83} : A_{27}. \bar{A}_{20}$;

$A_{84} : A_{12}. \bar{A}_{20}$; $A_{85} : A_{13}. \bar{A}_{20}$; $A_{86} : A_{14}. \bar{A}_{20}$;

$A_{87} : A_{15}. \bar{A}_{20}$; $A_{88} : A_{8}. \bar{A}_{20}$; $A_{89} : A_{9}. \bar{A}_{20}$;

$A_{90} : A_{10}. \bar{A}_{20}$; $A_{91} : A_{11}. \bar{A}_{20}$; $A_{92} : A_{24}. \bar{A}_{19}$;

$A_{93} : A_{25}. \bar{A}_{19}$; $A_{94} : A_{26}. \bar{A}_{19}$; $A_{95} : A_{27}. \bar{A}_{19}$;

$A_{96} : A_{12}. \bar{A}_{19}$; $A_{97} : A_{13}. \bar{A}_{19}$; $A_{98} : A_{14}. \bar{A}_{19}$;

A_{99} : A_{15}. $\bar{A}_{19}$; A_{100} : A_{8}. $\bar{A}_{19}$; A_{101} : A_{9}. $\bar{A}_{19}$;

A_{102} : A_{10}. $\bar{A}_{19}$; A_{103} : A_{11}. $\bar{A}_{19}$; A_{104} : A_{44}. $\bar{A}_{20}$;

A_{105} : A_{45}. $\bar{A}_{20}$; A[illegible] $\bar{A}_{20}$; A_{107} : A_{47}. $\bar{A}_{20}$;

A_{108} : A_{48}. $\bar{A}_{20}$; A_{109} : A_{49}. $\bar{A}_{20}$; A_{110} : A_{50}. $\bar{A}_{20}$;

A_{111} : A_{51}. $\bar{A}_{20}$; A_{112} : A_{52}. $\bar{A}_{20}$; A_{113} : A_{53}. $\bar{A}_{20}$;

A_{114} : A_{54}. $\bar{A}_{20}$; A_{115} : A_{55}. $\bar{A}_{20}$; A_{116} : A_{44}. $\bar{A}_{19}$;

A_{117} : A_{45}. $\bar{A}_{19}$; A_{118} : A_{46}. $\bar{A}_{19}$: A_{119} : A_{47}. $\bar{A}_{19}$;

A_{120} : A_{48}. A_{19} ; A_{121} : A_{49}. $\bar{A}_{19}$; A_{122} : A_{50}. $\bar{A}_{19}$;

A_{123} : A_{51}. $\bar{A}_{19}$; A_{124} : A_{52}. $\bar{A}_{19}$; A_{125} : A_{53}. $\bar{A}_{19}$;

A_{126} : A_{54}. $\bar{A}_{19}$; A_{127} : A_{55}. $\bar{A}_{19}$

(II) B-factors

B_1, B_2, B_3, B_4 (as stated in § 4.10)

(III) C-factors

C_1, C_2, C_3, C_4 (as stated in § 4.10)

Thus in the third classification we get in all 127 x 4 x 4 = 2032 combinations. Udyotakara says, ते एते हेतवो हेत्वाभासाश्च असङ्कीर्णविशेषणविशेष्यभेदेन द्वे सहस्रे द्वात्रिंशके ।[15]

Vācaspatimiśra explains, तदेवम् एभिः त्रिभिः चतुरशीतैः शतत्रयैः द्विपञ्चाशच्छतोत्तरं सहस्रम् एकं द्वात्रिंशदुत्तरया अष्टशत्या सहस्त्रमेकं चतुरशीतिसहिता च नवशती, तदेवं प्रकृतिभूताष्टचत्वारिंशतप्रक्षेपेण सम्पिण्डितं द्वात्रिंशे द्वे सहस्रे भवतः इति ।[16]

Vācaspatimiśra counts these combinations in the following manner. We have got 24 A- factors based on each of $\bar{A}_{18}$ (*anyathāsiddha*), $\bar{A}_{19}$ (*viruddhaviśeṣya*) and $\bar{A}_{20}$ (*viruddhaviśeṣaṇa*). 24 A- factors combined with B-factors and C-factors give 24 x 4 x 4 = 384 (*caturaśītaṁ śatatrayam*) combinations. Thus the total combinations based on $\bar{A}_{18}$, $\bar{A}_{19}$ and $\bar{A}_{20}$ are 384 x 3 = 1152 (*dvipañcāśacchatottaraṁ sahasram ekam*). The total A- factors based on $\bar{A}_1$ to $\bar{A}_{17}$ are 52 in number. These combined with B- factors and

C-factors give 52 x 4 x 4 = 832 (*dvātriṁśaduttarā aṣṭaśatī*) combinations. Thus the combinations of the second and the third classification are in all 1152 + 832 = 1984 (*sahasram ekam caturaśītisahitā ca navaśatī*) Then adding 48 combinations of the first classification to this figure, we get the total of 2032 combinations. Out of all these, only five combinations (mentioned in the first classification) refer to good *hetus*.

§ 4.15

Udyotakara's language while giving the list of these *hetus* and *hetvābhāsas*, sometimes becomes obscure and we have to take help from Vācaspatimiśra's clarification of the same. Udyotakara refers to his own work namely *Hetvābhāsavārtika*, which he prescribes for the examples of the various fallacies that Udyotakara refers to in *NV*, but leaves unexemplified.[17] However, the work is not available.

Some elementary states of *hetu* expressed by Udyotakara had already been suggested by Diṅnāga. For instance, Diṅnāga suggested the fallacies based on the dubitability of *hetu*,[18] which are elaborately considered by Udyotakara as the combinations of $\bar{A}_5$, $\bar{A}_6$, $\bar{A}_{11}$, $\bar{A}_{12}$. Diṅnāga had discussed the example of inference, where *hetu* exists somewhereelse than in *pakṣa*- subject.[19] Following such examples, Udyotakara coined the fallacies based on $\bar{A}_7$ and $\bar{A}_8$. Fallacies based on non-acceptance by either of the debaters or by both of the debaters (i.e. the combinations of $\bar{A}_9$, $\bar{A}_{10}$ .... $\bar{A}_{16}$) are also originated in Diṅnāga.[20] However, Diṅnāga did not mention the combination A_1 B_4 C_4 (which refers to *anupasaṁhārī*). The fallacies based on superfluity of *hetu* (combinations of $\bar{A}_3$ and $\bar{A}_4$) seem to be a new discovery by Udyotakara, though fallacy of superfluous thesis was mentioned by Śaṅkarasvāmin.[21] Of course, whether fallacies can be genuinely based on superfluity is a problem.[22]

§ 4.16

The fallacious inferences where '*hetu* can be explained otherwise' (*anyathasiddha* i.e. $\bar{A}_{18}$) seem to be first introduced by Udyotakara.

The *hetu* is called *anyathāsiddha* when the existence of *hetu* is capable of full explanation without reference to *sādhya*. This is in fact a way of showing the possibility of *vyabhicāra*, showing that *hetu* may exist without *sādhya*. The example of this fallacy discussed by Udyotakara is as follows,

'Shadow is a substance

because it moves'

Here, Udyotakara shows that the movement of shadow can be explained without assuming that shadow is a substance. The shadow, as he explains, moves not due to its substanciality but due to the movement of some substance which covers the light.[23]

This fallacy introduced by Udyotakara makes us think of some notion of 'necessary'[24] connection between *hetu* and *sādhya*, because if *hetu* 'necessarily' implies *sādhya*, then alone we can say that *hetu* cannot be explained fully without reference to *sādhya*. However, such a notion of 'necessary' connection is never positively noticed by Udyotakara. It will be seen in the next chapter that Dharmakīrti, who came after Udyotakara, based his theory of inference on the basis of *avinābhāva* as the 'necessary' connection

Notes

1. See, for instance, *HIL* p. 130; *NNTI* pp. 90-91; *FAL* p. 189.
2. *NV* p. 45.
3. *NV* p. 131.
4. All see असन्दिग्धमिति सजातीयाविनाभावि । *NV* p. 49.
5. Diṅnāga made an exception to the rule that *vipakṣa* must exist (See § 3.9). However, he did not approve of *kevalānvayi* inference, as explicitly as Udyotakara did.
6. This classification appears in Udyotakara's commentary of '*trividham*' in Gautama's definition of *anumāna* (*NS* 1.1.5). But what Udyotakara actually classified into three classes (namely *anvayī*, *vyatirekī* and *anvayavyatirekī*) is *hetu* and not *anumāna*. However, correspondingly we can classify *anumāna* and perhaps *vyāpti* also. Udyotakara also calls *avīta hetu* as *vyatirekī hetu*. (See

NV p. 123). Probably the other two types of *hetu* would come under *vīta hetu*.

7. See § 3.10.

7a. This kind of formal *vyāpti* is generally not mentioned by Indian logicians; but Dharmakīrti hints at it. See § 5.8.

8. कियता पुनर्भेदेनायं हेतुहेत्वाभासः प्रवृत्त इति । कालपुरुषवस्तुभेदानुविधानेन अपरिसङ्ख्येयः । *NV* p. 163.

9. Here and elsewhere in this chapter, the 'state of ***hetu***' does not mean the existence of ***hetu***. Even the non-existence of the expressed ***hetu*** would be regarded in this context, as a state of ***hetu***.

10. This state of *hetu* can be true only in case of the inferences where *pakṣa*-subjects are many. Thus the 'first step' (See § 3.8) in such inference would be of the form 'Ha. Hb. Hc.....' and the thesis (or conclusion) would be of the form 'Sa. Sb. Sc....'. There '***hetu*** exists in a part of *pakṣa*' means some conjuncts (and not all conjuncts) in the 'first step' are false.

11. For *hetucakra* of Diṅnāga see *NM* p. 28.

11a. Here onwards I mean by defective factor, an actual state of ***hetu***, which refers to a fallacy of ***hetu***, due to the incoherence between the asserted states of *hetu* and the actual state of ***hetu***.

12. The terms 'subject of ***hetu***' and 'predicate of ***hetu***' are significant when the *hetu* itself is a complex predicate. Though ***hetu*** itself is supposed to be the predicate of *pakṣa*-subject, here the 'subject of ***hetu***' does not mean *pakṣa*-subject. The terms 'subject of ***hetu***' and 'predicate of ***hetu***' are applied to the cases of double predication; that is to say, when the ***hetu*** consists of one subject and one predicate out of which predicate is predicated of the subject and subject is predicated of the *pakṣa*- subject. The 'first step' (See § 3.8) in such a case would be of the form F Ha, which means 'a has H, which has F'.

13. *NV* pp. 163-4.

14. *NVT* Part I, p. 331.

15. *NV* pp. 168-9.

16. *NVT* Part I, p. 335.

17. एषां तूदाहरणानि हेत्वाभासवार्तिके द्रष्टव्यानि स्वयं चाभ्यूह्यानि । *NV* p. 168.

18. See § 3.23.

19. The example is 'Sound is non-eternal, because the other categories like action would be equally eternal'. See *NM* p. 19.

20. See § 3.25.

21. See § 3.19.

22. See § 8.48.
23. कथमन्यथा ? आवरणसन्तानादसन्निधिसन्तानस्तेजस इति । आवरके द्रव्ये सर्पति तेजसोऽसन्निधिविशिष्टं यदुपलभ्यते तत्तु छायेत्युच्यते । *NV* p. 175.
24. The 'necessity' here should be regarded as empirical necessity and not as logical one.

Chapter V

The Nature of Anumāna and Sādhanābhāsa According to Dharmakīrti

Part I

§ 5.1

We have seen that Diṅnāga's form of *anumāna* was that of a strong analogical reasoning. Though he systematised the theory of *anumāna* which was given by earlier Nyāya, Vaiśeṣika and Buddhist systems, he did not consider the problem of formal validity; he only considered the problem of truth of the premises and conclusion of an argument.

Invalid inference cannot guarantee the truth of its conclusion, though its premises are known to be true. If on the other hand the inference is valid and it has true and only true premises, we can certainly be sure about the truth of its conclusion. Such an inference which is valid and also has only true premises is regarded as *pramāṇa* by classical Indian logicians. Let us call such inference a sound inference. The task before Dharmakīrti was to state and justify the frame of such a sound inference. The problem before him was two-fold,

1) How to have a valid form of inference?

2) How to be sure about the truth of the premises?

In this first part of the chapter we are going to consider and examine in brief, the efforts made by Dharmakīrti to solve these two problems.

§ 5.2

It is observed that Diṅnāga's premises of *parārthānumāna* cannot entail the conclusion.[1] This was the case mainly because the scope of *vyāpti* according to Diṅnāga was restricted to the field of *sapakṣa* and *vipakṣa*, that is to say, to the field consisting of the observed cases similar to the thesis and dissimilar to the thesis. But the scope of *vyāpti* according to him was not pervasive enough to include in itself the case of the thesis. Dharmakīrti on the other hand held that *vyāpti*, namely the dependence-relation between *hetu* and *sādhya* must apply to all cases, observed or unobserved. Dharmakīrti uses the term *sarvopasaṁhāra*[2] to express the idea of all-pervasiveness.

§ 5.3

Some of the examples of *vyāpti*-statement, given by Dharmakīrti are as follows,

1) यदुपलब्धिलक्षणप्राप्तं सन्नोपलभ्यते सोऽसद्व्यवहारविषयः सिद्धः ।3

(That which is otherwise capable of being obtained but is not obtained is *proved to be* the object, which can be called non-being)

2) यत् सत् तत् सर्वमनित्यम् ।[4] (*Everything* that is real is impermanent)

3) यत् सद् उपलब्धिलक्षणप्राप्तं तद् उपलभ्यते एव ।[5] (That which is capable of being obtained, and is present, gets actually obtained with *certainty*.)

4) अनित्यत्वाभावे कृतकत्वासम्भवः ।[6] (If something is not impermanent then it is *impossible* that it can be a product).

5) यदुत्पत्तिमत् तदनित्यम् ।[7] (That which has origination is impermanent).

The first four examples of vyāpti- statement indicate that the relation between *hetu* and *sādhya* should be already proved (*siddha*), should be 'universal'[7a] ('*sarvam*') should be certain ('*eva*') and its negation should be impossible ('*asambhava*') These terms prohibit the possibility that in a particular case of true *vyāpti*-statement, *hetu* may be existing without *sādhya*. Though the fifth example of *vyāpti*-statement mentioned above is silent about this kind of explicit indication, it should be distinguished from the type of *vyāpti*-statement accepted by Diṅnāga.

We have seen (in § 3.11) that in Diṅnāga's statements of *'vyāpti'*, the term *dṛṣṭam* is significant. The import of Diṅnāga's use of the term *dṛṣṭam* lies in his presupposition that we base our inference on the observed relation between *hetu* and *sādhya*. The term *dṛṣṭam* is significant in relation with the term *dṛṣṭānta*. Here we have to note the distinction between *dṛṣṭānta* and *siddhānta*. The universal statement of *vyāpti* can be called a kind of *siddhānta* but it cannot be subsumed under *dṛṣṭānta*. Diṅnāga included his statement of *vyāpti* in the *dṛṣṭānta*-element of *parārthānumāna* because his *vyāpti*-statements were reducible to the conjunctions of observation-statements. Dharmakīrti, however, clearly distinguishes between a *vyāpti*-statement and a *dṛṣṭānta*-expression. Many a time he uses the word *dṛṣṭa* while mentioning the particular instance (*dṛṣṭānta*) but never uses it in the *vyāpti*-statement as Diṅnāga does. See, for instance, Dharmakīrti's use of *siddha* and *dṛṣṭa*,

यदुपलब्धिलक्षणप्राप्तं सन्नोपलभ्यते सः असद्व्यवहारविषयः सिद्धः यथान्यः कश्चिद् दृष्टः शशविषाणादिः ।[8]

Here *siddhaḥ* is the description of 'universal' *vyāpti* and *dṛṣṭaḥ* is the description of the particular instance. Here *dṛṣṭaḥ* does not mean 'perceived' in the strict sense of the term, but means 'noticed as an instance'. (The 'instance' may be even a ficticious object like the horn of a hare).

This attitude of Dharmakīrti clearly shows that Dharmakīrti's notion of *vyāpti* was that of a 'universal' and 'necessary' relation between *hetu* and *sādhya*.

§ 5.4

Dharmakīrti being a grand-disciple of Diṅnāga, tries not to be in conflict with Diṅnāga's views. However, on some occasions, Dharmakīrti cannot help. When, for instance, Dharmakīrti finds that the fallacy called *viruddhāvyabhicāri* mentioned by Diṅnāga need not be mentioned as a distinct fallacy when we are considering the frame of sound inference, Dharmakīrti says that he is in deep distress because then he has to discover his Guru's mistake.[9] This occasion in *Hetubindu* suggests that in general Dharmakīrti preferred not discover Diṅnāga's weaknesses as a logician, and tried to interpret in the manner which would protect Diṅnāga. I have shown in § 3.5 how Dharmakīrti tried to show that Diṅnāga had the idea of necessary relation between *hetu* and *sādhya* when Diṅnāga talked of three CEs of *hetu*. I also opined that Dharmakīrti has misinterpreted Diṅnāga. It should be noticed here that though Dharmakīrti as a follower of Diṅnāga repeated Diṅnāga's idea of three CEs of *hetu*, he gave new meanings to the second and the third CE. He used the words *eva* and *niścita*[10] with the special intention that the second CE would ultimately mean the 'universal' *anvaya-vyāpti* and not just *sapakṣānugama*. Similarly the third CE would ultimately mean the 'universal' *vyatirekavyāpti* and not just *vipakṣavyatireka*. He clarifies his intention when he interprets the word *niścita* in terms of *pratibandha*[11] (= the relation of '*dependence*' between *hetu* and *sādhya*). At most occasions in *Hetubindu*, Dharmakīrti prefers to use the term *anvaya* and *vyatireka* to connote the two aspects of *vyāpti*, rather than *sapakṣasattva* and *vipakṣāsattva*.

§ 5.5

Dharmakīrti thus succeeded in having a valid form of *parārthānumāna* by introducing universal statement of *vyāpti* as one of the premises. At this stage, however, the second problem before Dharmakīrti, namely, how to be sure about the truth of the premises becomes more complicated. For the earlier logicians, for whom *anumāna* was just analogical reasoning, the premises were the observation statements and thus were empirically verifiable. Thus the justification or falsification of these statements was to be made by perception. In Dharmakīrti's model of *anumāna* the premise which stated *pakṣasattva* (i.e. the characteristic of *hetu* that it exists in *pakṣa*-case) was an observation-statement. But *vyāpti*-statement which expressed either *anvaya-vyāpti* or *vyatireka-vyāpti*, was not an observation-statement. Dharmakīrti states firmly that *vyāpti* cannot be established either by observation or by non-observation.[12] The reason seems to be simple. One can know through perception, only the particular relation (like *saṁyoga*) between particular *hetu* and particular *sādhya*. But the universal relation between 'universal *hetu*' and 'universal *sādhya*' cannot be known through mere perception.

If it is granted that according to Dharmakīrti any universal is known only through *anumāna*,[12a] then we will have to suppose that, for Dharmakīrti, *vyāpti* being a universal relation is known only through *anumāna*. Here we have to face a greater difficulty because *anumāna* which proves this *vyāpti*(1) needs the knowledge of some other *vyāpti*(2) through which *vyāpti*(1) can be proved. And this leads to infinite regress. Thus we can neither accept that *vyāpti* in Dharmakīrti's concept of inference, can be known through *anumāna*, nor can we accept that it can be known through perception. Dharmakīrti's system rules out the third possibility because he accepts only two *pramāṇas*. Thus the problem before Dharmakīrti, namely, how to be sure about the truth of the premises, remains ultimately unanswered, because the *vyāpti*-statement which is one of the premises of *anumāna* cannot be known or proved to be true.

§ 5.6

Dharmakīrti's one of the main theses was that the relation of *vyāpti* should not be only a universal relation, but also a necessary one. Thus the universal statement which expresses the inductive generalisation made from the observed co-existence and co-absence, will not be properly called a *vyāpti*-statement unless it is supported by some necessary connection between those relata. The possibility of *vyabhicāra* can be thought of where there is merely observed constant co-existence between *hetu* and *sādhya*. But the possibility of *vyabhicāra* cannot be consistently thought of when there is the necessary relation between *hetu* and *sādhya*. The necessary relation according to Dharmakīrti, may be of either of the two kinds,

1) The relation of having the same reference (*tādātmya*)

2) The causal relation between *hetu* and *sādhya* (where *sādhya* is the cause of *hetu*). (*tadutpatti*)

It can be seen from the above types of 'necessary relation', that the 'necessary relation' between *hetu* and *sādhya* need not always be logically or conceptually necessary relation; it can even be empirically necessary relation. That is to say, the statement of *vyāpti*, for being a 'necessary' statement, need not be an analytic statement. In the case of *tādātmya* it may be claimed that *sādhya* follows analytically from *hetu*; it may be said for instance that the *vyāpti*-statement like 'Every Sissoo-tree is a tree' (In Sanskrit,'यत्र यत्र शिंशपात्वं तत्र तत्र वृक्षत्वम्') is an analytic statement (though Dharmakīrti would not regard it as an analytic statement but as a synthetic generalisation[13]). But same thing is not true in the case of tadutpatti. The vyāpti-statement like 'Whatever has smoke has fire' (which is based on tadutpatti) cannot be regarded as an analytic statement. Thus the necessity here is not analyticity.

Dharmakīrti calls this relation *pratibandha*, which means binding or dependence. This dependence can be expressed through the logical connective 'implies' with the reservation that here 'implies' does not refer to 'any' sort of material implication but to the sort of material implication which involves the necessity of either *tādātmya*-type or *tadutpatti*-type.

§ 5.7

Dharmakīrti also seems to be aware of the distinction between material implication and formal implication. The term *pratibandha* used by Dharmakīrti connotes a kind of material relation and not a formal relation. Even in the case of the inference based on *tādātmya* between *śiṁśapātva* as *hetu* and *vṛkṣatva* as *sādhya*, Dharmakīrti would say, that the *hetu-sādhya*-relation is not grasped just by understanding the concepts *śiṁśapā* and *vṛkṣa*, but is grasped by observing the things called *śiṁśapā* and *vṛkṣa*. In this sense, Dharmakīrti may even hold that 'Every *śiṁśapā* is a tree' is not an analytic statement. Though it is true that when we refer to a thing *śiṁśapā*, we also refer at the same time to a thing *vṛkṣa*; Dharmakīrti does not appear to believe, as a conceptual atomist does, that some concepts are contained in some other concepts, say for instance, that *śiṁśapātva* contains (involves, consists of) *vṛkṣatva*. Thus according to Dharmakīrti, the relation amongst the concepts *śiṁśapā* and *vṛkṣa* is not given to us with the concept *śiṁśapā*, but it is to be known by observing the actual denotations of the words.

Here we have to distinguish clearly between the two terms used by Dharmakīrti namely *pratibandha* and *sāmarthya*. It seems that *pratibandha* stands for a material relation of dependence amongst the things or properties, while *sāmarthya* stands for a formal or semantic relation between two or more statements. A statement of *pratibandha* is formulated by observing the denotations of the terms to be used in the statement. The knowledge of denotations, however, is not needed for validly deriving one statement from another. The knowledge of connotations suffices. We see that when Dharmakīrti discusses the nature of *vyāpti*, which stands for a material relation, he talks in terms of *pratibandha*. Even while discussing the *tādātmya* kind of *vyāpti*, Dharmakīrti is not seen using the term *sāmarthya*. He is never found saying that, given that something is *śiṁśapā*, it follows by the semantic force of the statement that it must be *vṛkṣa*. There he talks in terms of *svabhāva*. For example he would say that it is the *svabhāva* of *śiṁśapā* that it is a tree. But *svabhāva* again stands for the material intrinsic nature of a thing, and not for the connotation of a term or a sentence. For conveying the idea of 'the semantic force of a statement by which it implies another statement', Dharmakīrti uses the expressions like *'sāmarthyāt eva bhavati'*[14], *'sāmarthyāt .... ākṣipati'*[15], where *sāmarthya* means the force of meaning (connotation) of the given statement or statements. Thus the concept of *pratibandha* with the concepts of *tādātmya* and *tadutpatti* is concerned with ontological facts, while the concept of *sāmarthya* is concerned with logico-linguistic facts. Dharmakīrti talks in terms of *sāmarthya* when he wants to say that the conclusion of *parārthānumāna* follows logically from the premises[16], or that from the statement of *anvayavyāpti*, *vyatirekavyāpti* can be logically derived and vice-versa.[17]

§ 5.8

We find that in Diṅnāga's view of *parārthānumāna*, the statement of *anvaya* (= the second step of P_1) and the statement of *vyatireka* (= the third step of P_1) do not imply each other.[18] For Dharmakīrti, however, the statement of *anvaya* or the statement of *vyatireka* is implied by the single statement of *pratibandha*. Such a statement of *pratibandha* could be either of the form,

For all x if x has *hetu* then x has *sadhya*.

or of the form,

For all x if x does not have *sādhya* then x does not have *hetu*.

Here it is clear that the two formulations are logically equivalent with each other.[18a]

§ 5.9

Though Dharmakīrti generally adds the instanciation-part to the statement of *vyapti*, he also says that the separate mention of *dṛṣṭānta* as an element of *parārthānumāna* is redundant.[19]

It seems that according to Dharmakīrti the function of instanciation-part is only to give a hint for the proper understanding of the *vyāpti*-statement. But the instanciation-part is neither logically necessary nor logically sufficient for proving *vyāpti*. We also find some examples of sound reasoning given by Dharmakīrti, which do not include the instanciation-part. See for instance,

1) If something is not impermanent, then it cannot be real (or a product or a creation). But the sound is real (or a product or a creation).[20]

2) Anywhere, if fire is not there, smoke cannot be there. But here there is smoke.[21]

Sometimes Dharmakīrti uses 'non- existents' like the horn of a hare, as instances in sound reasoning.[22] This shows that according to Dharmakīrti, the instanciation of universal *vyāpti*-statement need not have existencial import. The instanciation may be non-existencial. Diṅnāga had referred to the instance like space (ākāśa)[23] which was existent according to the arguer (Vaiśeṣika) though non-existent according to the opponent (Buddhist). But Dharmakīrti's instance of the horn of a hare is existent for nobody.

The views of Diṅnāga and Dharmakīrti regarding instanciation are reflected in their views about some fallacies also. If in an argument, instanciation is lacking, Diṅnāga would regard it as a fallacy called *nyūna* (insufficient reasoning).[24] But as according to Dharmakīrti the instanciation has no function over and above giving a hint for the knowledge of *vyāpti*, the argument whicl contains *vyāpti*-statement as a premise, but lacks its instanciation, would not be called a case of insufficient reasoning according to Dharmakīrti.

According to Dharmakīrti, as according to Diṅnāga, the instanciations are of the form[25]

d (which is other than *pakṣa*- subject) has both *hetu* and *sādhya*,

(positive instance)

d̄ (which is other than *pakṣa*-subject) has neither *sādhya* nor *hetu*.

(negative instance)

In both the instanciations the clause of the form 'which is other than *pakṣa*-subject' is presupposed as necessary for indicating that the subject of instance is not the same as the *pakṣa*-subject.

§ 5.10

Dharmakīrti also says that the mention of the conclusion in the full argument is not needed because the conclusion by virtue of the semantic force of the premises (*sāmarthyād eva*)[26] follows from them. It seems that in the inferencial model proposed by Dharmakīrti the thesis is already stated before the argument, in the form of *pakṣa*, and the argument follows as the reasoning supplied to prove that *pakṣa*. According to Dharmakīrti *pakṣa* was not included in the actual body of the proof (*sādhana*) but was kept outside, when according to Diṅnāga *pakṣa* was to be included as a part of proof. We have seen that Diṅnāga's form of argumentation was like the following

P Because R

where P stands for *pakṣa* and R stands for the set of premises, when the popular form of argument is like the following,

R Therefore P

Here P stands for *nigamana* rather than for *pakṣa*. It may be noted here that the logical status of both the forms is the same (though the epistemological status differs). Dharmakīrti's form of argumentation is, however, like the following,

R

where R stands for the set of premises. This is the form of the enthymeme of the third order.[27]

§ 5.11

Dharmakīrti, however, does not seem to be fully opposed to the status of *pakṣa* in the context of *parārthānumāna*. Two statements of Dharmakīrti should be read carefully in this context.

1) द्वयोरप्यनयोः प्रयोगे नावश्यं पक्षनिर्देशः ।[28]

(Meaning : In both these types of argument mentioning *pakṣa* is not needed)

2) कीदृशः पुनः पक्ष इति निर्देश्यः ? स्वरूपेणैव स्वयम् इष्टः अनिराकृतः पक्षः इति ।[29]

(Meaning : What should be mentioned as *pakṣa*? *Pakṣa* should be that which is intended to be proved by the debater himself and which is not already falsified).

The two statements suggest that though Dharmakīrti must have allowed the statement of *pakṣa* to be presented before the proof (*sādhana*), he considers *pakṣa* to be outside the body of proof. Thus, there does not seem to be any basic difference of opinion between Dharmakīrti and the other logicians regarding the problem whether *pakṣa* should be mentioned in course of arguing or not. Therefore, when we come to present Dharmakīrti's forms of sound inference, we include *pakṣa* also in the form of argument.

§ 5.12

The forms of sound *parārthānumāna* Dharmakīrti had in mind can be presented by using the stock-example of inference of impermanence of sound in the following way -

P_3: Sound is impermanent

because sound is a product

Whatever is a product is impermanent.

For example pot (which is other than sound) is

both a product and impermanent.[29a]

P_4 : Sound is impermanent

because sound is a product.

Whatever is not impermanent is not a product.

For example ether (which is other than sound) is

neither impermanent nor a product.

In P_3 we have stated and instanciated *anvaya-vyāpti* whereas in P_4 we have stated and instanciated *vyatireka-vyāpti*. Dharmakīrti calls P_3 *sādharmyavat prayoga* and P_4 *vaidharmyavat prayoga.*

As we have seen in § 5.9, instanciation-part can be omitted. Thus we have,

P_5: Sound is impermanent

because sound is a product.

Whatever is a product is impermanent.

P_6: Sound is impermanent

because sound is a product.

Whatever is not impermanent is not a product.

Thus we see that according to Dharmakīrti we can have two major modes of arguing for the same thesis from the same *hetu*. These modes are Modus Ponuns and Modus Tollens.

In *Vādanyāya*, Dharmakīrti emphasises the point that the order of the premises need not be fixed.[30] Thus we get different forms by changing the order of the premises. For instance,

P_7 : Sound is impermanent

because whatever is a product is impermanent

and sound is a product.

P_8 : Sound is impermanent

because whatever is not impermanent is not a product

but sound is a product.

Part II
§ 5.13

Dharmakīrti's treatment of *sādhanābhāsa* resembles Diṅnāga's account, but also differs from it in some very important respects. So instead of giving the complete account of fallacies discussed by Dharmakīrti, we shall discuss only the similarities and the differences between the two accounts.

Diṅnāga thought that the inference had three elements namely *pakṣa, hetu* and *dṛṣṭānta*. But as we saw in Part I, Dharmakīrti does not give much importance to *pakṣa* or *dṛṣtanta*. He therefore gives greater importance to *hetu* and discusses *hetvābhāsas* with great details. He uses to terms *hetvābhāsa* and *sādhanābhāsa* almost synonymously.

§ 5.14

Diṅnāga considered five types of *pakṣābhāsa*;[31] Dharmakīrti recognised only four. Dharmakīrti does not recognise *āgamaviruddha*. Perhaps Dharmakīrti accepted Udyotakara's criticism[32] of *āgamaviruddha*, and thought that *āgamaviruddha* could be included in *anumānaviruddha*. In spite of this difference, Dharmakīrti's classification of *pakṣābhāsa* is essentially the same as that of Diṅnāga.

Dharmakīrti does not regard *pakṣa* as a part of *sādhana* (proof). So he is not prepared to include *pakṣābhāsa* in *sādhanābhāsa*. I however think that the thesis or conclusion should be regarded as a part of argument.[33] It is true that, as Dharmakīrti points out, the thesis, even when it is not made explicit, logically follows from the sufficient premises. But it is better if one expresses the intended thesis while arguing, because many conclusions might follow from the same set of premises, and the intended conclusion might not be specified otherwise. *Pakṣa* should thus be regarded as an implicit (in case of ethymeme) or explicit part of the argument and accordingly *pakṣābhāsa* should be included in *sādhanābhāsa*.

§ 5.15

According to Dharmakīrti, an argument is fallacious under any of the following conditions.[34]

1) Some CE of *hetu* is not expressed in the premises.

2) All the CEs are expressed, but some CE is false (non-acceptable) according to some debater.

3) All the CEs are expressed, but some CE is dubitable according to some debater.

It can be seen therefore that the basic principles of fallacies according to Dharmakīrti are insufficiency (invalidity), falsity, dubitability and non-acceptance.

The non-expression of any CE of *hetu* in the premises was called by Dharmakīrti as *nyūna* (insufficient reasoning). Though Diṅnāga had recognised the fallacy called *nyūna*, his *nyūna* cannot be properly called a formal fallacy, because Diṅnāga's form of good inference itself was not a valid form.[35] In Dharmakīrti's form of inference, *nyūna* attains greater significance from the point of view of formal logic. The premises of inference which Dharmakīrti discussed were the necessary and sufficient premises of a sound *anumāna*. Thus the inferences of the form,

(1) Sound is a product

Therefore sound is impermanent.

(2) Whatever is a product is impermanent

Therefore sound is impermanent.

which are the cases of *nyūna*, are invalid inferences. Thus *nyūna* discussed by Dharmakīrti is a formal fallacy in which premises are insufficient to entail the conclusion.

§ 5.16

While explaining the other *hetvābhāsas* discussed by Dharmakīrti, we use the same method as we used in explaining Diṅnāga's account. Let us put before us the form of *parārthānumāna* accepted by Dharmakīrti as a form of sound inference. We can consider, for instance the example of the form P_5, which runs,

Thesis: Sound is impermanent.

Because,

First step : Sound is a product

Second step : Whatever is a product is impermanent.

The first step in this form is the same as that in Diṅnāga's form (P_1 or P_2). Thus the fallacies of *hetu* based on falsity/dubitability/non-acceptance of the first step would be the same, both for Diṅnāga and Dharmakīrti. Dharmakīrti's fallacies of the first step of P_5 may be enumerated as follows.

1) The first step not accepted by both the debaters.

(*ubhayāsiddha*, Also see § 3.25)

2) The first step not accepted by the opponent.

(*prativādyasiddha*)

3) The first step not accepted by the arguer himself.

(*svayaṁ vādino' siddha*)

2) and 3) are the subtypes of Diṅnāga's *anyatarāsiddha* (See § 3.25)

4) Hetu is doubted to exist.

(*svayaṁ sandigdhāsiddha*)

5) The *pakṣa*-subject is doubted to exist.

(*sandigdha-āśrayaṇāsiddha*)

4) and 5) are the subtypes of Diṅnāga's *sandigdhāsiddha* (See § 3.24)

6) The existence of *pakṣa*-subject is falsified.

(*āśrayaṇāsiddha*, also see § 3.23)

§ 5.17

The second step of Dharmakīrti's P_5 is very much different from that of Diṅnāga's P_1 or P_2. Dharmakīrti in fact uses a single step (second step of P_5) in place of Diṅnāga's two steps, the second and the third. Thus all the remaining fallacies of *hetu* according to Dharmakīrti are the fallacies of the second step of P_5. These are the fallacies based on falsity or dubitability of *vyāpti*-statement.

The second step of P_5 is the statement of *anvaya-vyāpti*, the form of which is as follows.

Whatever has *hetu*-predicate has *sādhya*- predicate.

The step is false if and only if there is something which has *hetu*-predicate but not *sādhya*-predicate.

This situation obtains in *sādhāraṇa anaikāntika* as also in *viruddha*. All the varieties of *sādhāraṇa anaikāntika* that we discussed in § 3.28 can amount to the negation of the second step of P5. Similarly both the varieties of *viruddha* that we discussed in § 3.29 also can amount to the negation of the second step of P5. We have also seen that the second step of P5 and the second step of P6 are equivalent with each other.[36] Therefore fallacies based on the falsity of *anvayavyāpti* and *vyatirekavyāpti* are the same.

§ 5.18

We cannot explain, however, the fallacy namely *asādhāraṇa anaikāntika* in the same manner, because in Dharmakīrti's premises of *parārthānumāna* we nowhere find included

"There is something (other than *pakṣa*-subject) which has both *hetu* and *sādhya*" which Diṅnāga accepts as a part of the second CE.[37] If *hetu* is neither predicated of *sapakṣa*-subjects nor of *vipakṣa*-subjects, it does not follow that the *vyāpti*-statement must be false.

However, Dharmakīrti recognises the fallacy namely *asādhāraṇa anaikāntika* on a different ground. According to him, though the non-availability of empirical evidence for *sapakṣa-sattva* does not falsify the *vyāpti*-statement, still it can show that the *vyāpti*-statement is uncertain. And thus the uncertainty or the dubitability of the *vyāpti*-statement can amount to the fallacy. Thus though both Diṅnāga and Dharmakīrti recognise the fallacy namely *asādhāraṇa anaikāntika*, their treatment of the fallacy differs. For Diṅnāga the fallacy consists in the falsity of the second step (See § 3.31); for Dharmakīrti the fallacy lies in the doubt about the truth of the second step (of P5). Dharmakīrti describes this fallacy as one which is discovered by raising doubt about both *anvaya* and *vyatireka* between *hetu* and *sādhya*.[38]

§ 5.19

Dharmakīrti also discusses another variety of fallacy in which *vyatireka*, but not *anvaya* is doubted.[39] Here *anvaya* and *vyatireka* should be taken in the sense of *sapakṣasattva* and *vipakṣāsattva* respectively. Here the intention of Dharmakīrti is to suggest that if either of *sapakṣasattva* and *vipakṣāsattva* is dubitable, the universal *vyāpti* also is dubitable.

The example runs, "Mr. X is not omniscient because he speaks".

In such examples one can find instances where *hetu* and *sādhya* co-exist, but can anyone get a case where *hetu* and *sādhya* are co-absent? For a doubt can always be raised regarding somebody's omniscience. While explaining the example Dharmakīrti suggests that if the *vyāpti*-statement is true, *vyabhicāra* must be impossible. But here, though the case of *vyabhicāra* (namely a speaking person who is omniscient) is not clearly discovered by us, it cannot be shown to be an impossibility. Between 'being omniscient' and 'the act of speaking' there is neither empirical inconsistency nor logical contradiction.[40]

This fallacious inference is similar to a *kevalānvayi*-type of inference, because it is based on merely the positive analogy between Mr. X and the non- omniscient persons, and not based on the negative analogy between Mr. X and some omniscient person. This suggests that even the inference based on *anvaya* alone may be sometimes fallacious, if the 'necessary' *vyāpti*-statement is dubitable.

§ 5.20

Dharmakīrti's treatment of the fallacy called *viruddha*, is basically traditional. That is to say, he does not differ from Diṅnāga as far as the basic nature is concerned. According to both Diṅnāga and Dharmakīrti the *hetu* which is fallacious due to *viruddha*, is predicated of at least some *vipakṣa*-subjects but not predicated of any *sapakṣa*-subejct.

It seems that Dharmakirti could have defined *viruddha* so as to suit to his theory of sound inference. Accordingly the situation of viruddha would take the form,

Whatever has *hetu*-predicate, lacks *sādhya*-predicate.

§ 5.21

Dharmakīrti discusses *iṣṭavighātakṛt* (the *hetu* that is contrary to the intended probandum)[41], a special type of *viruddha* recognised by Diṅnāga. The example runs,

"The sense-organs are useful for something else because they are combinations of atoms, just like bed, chair etc. which are combinations of atoms and are useful for the person who uses them."

Here the existence of soul is the intended probandum. But the probandum explicit in the argument is not soul, but 'something other than sense-organs'. When the arguer proves 'something', he proceeds to claim that it is nothing but the permanent soul.

The difficulty that is raised in the above inference is that the *hetu* namely 'being a combination of atoms' can at the most imply the existence of some other combination of atoms for the sake of which this combination is used. Thus the person (i.e. the body of the living person) also is a combination of atoms for the sake of which bed, chair etc. are used. Therefore the *hetu* is contrary to the implicit *sādhya* namely the soul (which is obviously not a combination).

Here the problem before Dharmakīrti is whether this type of *viruddha* should be regarded as a special type of *viruddha* or should be included in *viruddha* in general. Diṅnāga chose the first alternative while Dharmakīrti chose the second. Dharmakīrti argued that the intended (implicit) *sādhya* need not be distinguished from the stated (explicit) *sādhya* because both are expected to be proved.[42]

It should be noted here that one can discover a fallacy in a particular argument only by examining the explicit linguistic form and content of the argument. How can one be sure about the intended nature of *sādhya* unless it is explicit in the argument? Therefore, the kind of inference where only the implicit *sādhya* is contradicted by *hetu*, need not be considered as a strict case of *viruddha*. The above inference in which the explicit *sādhya* is 'something other than the sense-organs' and not 'soul', is not at all fallacious in the way in which Buddhists are arguing. The inference may be fallacious only when the explicit *sādhya* is 'something which is a non-combination'.

§ 5.22

The premises of argument described by Diṅnāga were not logically sufficient to entail the conclusion. Diṅnāga might have been aware of this fact when he introduced the fallacy called *viruddhāvyabhicāri*. (See § 3.34)

Diṅnāga regarded *viruddhāvyabhicāri* as a type of *anaikāntika*. However, *viruddhāvyabhicāri* was much different from the primary type of *anaikāntika*. The primary type of *anaikāntika* was only *sādhāraṇa anaikāntika* (so called later on), while Diṅnāga expanded the scope of the fallacy. Though Nyāya logicians after Diṅnāga accepted the expansion made by Diṅnāga, they took *viruddhāvyabhicāri* to be a distinct fallacy. This was because it has very peculiar characteristic not shared by any other type of *anaikāntika*. As Nyāya logicians following Diṅnāga accepted one to one correspondence betwee the CEs of *hetu* and the basic fallacies of *hetu*, they had to state a distinct CE of *hetu* corresponding to *viruddhāvyabhicāri*. Therefore they talked of *vivakṣitaikasaṅkhyatva* of *hetu*.[43]

Akṣapāda prefers to use the term *hetvābhāsa* when he talks of inferencial fallacies. He does not use the term *pakṣābhāsa* or *dṛṣṭāntābhāsa*. Following Akṣapāda the orthodox Nyāya logicians tried to convert every inferencial fallacy into a *hetvābhāsa*. They thus converted the fallacy of thesis into the fallacy of *hetu* and called it *bādhita*. They stated it as a distinct fallacy and formed a corresponding distinct CE of *hetu*, namely *abādhitaviṣayatva*.[44]

Nyāya logicians also thought that *hetu* can give rise to the knowledge of *sādhya*, only if *hetu* is already known to the one who infers. This characteristic of *hetu* was known as *jñātatva*[45] and Nyāya logicians supposed that this also should be added to the list of CEs of *hetu*. (It is not clear from *Hetubindu* whether Nyāya logicians also recognised the distinct fallacy of *hetu* corresponding to this CE of *hetu*.) All this movement of making additions in Diṅnāga's list of CEs of *hetu* was noticed and referred to in *Hetubindu* by Dharmakīrti. In *Hetubindu* Dharmakīrti mentioned the view of the opponents (most of whom were Naiyāyikas) who thought that the defining characteristics or the CEs of *hetu* are in all six and not just three. Thus the three CEs added by them were

1) *abādhitaviṣayatva*

2) vivakṣitaikasaṅkhyatva

3) jñātatva

We find that in the available ancient treatises of Nyāya, the first mention of five CEs of *hetu* appears in *Nyāyamañjarī* of Jayantabhaṭṭa. It seems that Naiyāyikas at the time of Dharmakīrti were advocating six CEs of *hetu*. Gradually they must have dropped the sixth CE namely *Jñātatva*, but at least from the time of Jayantabhaṭṭa, if not earlier, Naiyāyikas must have stuck to the number five.

§ 5.23

Dharmakīrti, however, rejects all the three added CEs. We will now see how and why these added CEs of *hetu* could not find any place in the theory of *anumāna* accepted by Dharmakīrti.

We have seen that Dharmakīrti tried to establish a form of sound inference, and for that he interpreted Diṅnāga's three CEs in a different way. For Dharmakīrti, if any *hetu* possesses these (newly interpreted) three CEs of *hetu*, then such a *hetu* cannot be fallacious. Thus as the addition in the list, of newly interpreted three CEs was unnecessary, the addition in the corresponding list of fallacies also was unnecessary.

An inference is fallacious due to *bādhita*, if either perception or inference or any authorised means of knowledge shows that *sādhya* is not predicated of *pakṣa*-subject. Thus if an inference is fallacious due to *bādhita* then the universal statement of *vyāpti* must be false. Because if *hetu* is predicated of the *pakṣa*- subject, but *sādhya* is not, then this is a clear case of *vyabhicāra* which falsifies the universal statement of the form 'Of whatever subject *hetu* is predicated, *sādhya* must be predicated'. *Bādhita* and the universal *vyāpti* cannot go together.[46] Thus *bādhita* is reducible to a clear case of *savyabicāra*, where *savyabhicāra* consists in the falsity of the *vyāpti*-statement. If so, *bādhita* need not be regarded as a distinct fallacy and correspondingly the mention of *abādhitaviṣayatva* as a distinct CE of *hetu* need not be accepted.

§ 5.24

Similarly when we consider *vivakṣitaikasaṅkhyatva* (or *asatpratipakṣatva*) as a distinct CE of *hetu*, it becomes clear that in the context of Dharmakīrti's newly interpreted three CEs, this additional CE need not be accepted.

If an inference is fallacious due to *viruddhāvyabhicāri* (or *satpratipakṣa*) then there must exist some alternative *hetu* which may imply the negation of *sādhya*. Thus the alternative inference gives the conclusion which contradicts the proposed conclusion. If it is granted that the proposed *hetu* is predicated of *pakṣa*-subject, and if the alternative *hetu* shows that *sādhya* may not be predicated of the *pakṣa*-subject, then it follows that there is a possibility of *vyabhicāra* (i.e. the situation where *hetu* may exist without *sādhya*) in *pakṣa*-case. Thus *viruddhāvyabhicāri* is reducible to a kind of *savyabhicāra*. Therefore *viruddhāvyabhicāri* or *satpratipakṣa* need not be recognised as a distinct fallacy of *hetu* and correspondingly the mention of *vivakṣitaikasaṅkhyatva* or *asatpratipakṣatva* as a distinct CE of *hetu* need not be accepted.

§ 5.25

Dharmakīrti suggests that though *hetu* leads us to the inferencial knowledge only if the *hetu* is known, still being known (*jñātatva*) is not CE (constitutive element) of *hetu*. We, according to Dharmakīrti, are here supposed to state the defining characteristics of *hetu* and the defining characteristics of *hetu* can be only those which are the characteristics of *hetu* itself (*ātmadharma*) and not the characteristics of something else (*pararūpa*).[47] Knowledge is not the property of *hetu* itself. Here Dharmakīrti seems to distinguish between constitutive characteristics and non-constitutive characteristics. He also seems to claim that only constitutive characteristics can be regarded as defining characteristics. He, for instance, argues that if non-constitutive characteristics of *hetu* are to be regarded as defining characteristics then the necessary conditions of good inference like 'the existence of some person who infers' or 'the existence of the object to be inferred' will also have to be regarded as the defining characteristics of sound *hetu*. But this is not intended by anybody.

§ 5.26

Though the instanciation-part is not logically necessary for the soundness of an argument, still for understanding the *vyāpti*-statement, the instanciation may be sometimes practically necessary. If the instanciation is included in the argument then it can be examined as a part of argument and thus we can recognise some fallacies of instance from which the argument should be free. Secondly, as has been stated in § 3.22, Diṅnāga mixed *hetu*-element with *dṛṣṭānta*-element and accordingly he discussed some fallacies concerned with *hetu* under the title *'dṛṣṭāntābhāsa'*. Such reasons seem to have made Dharmakīrti discuss *dṛṣṭāntābhāsas* in his theory of fallacies.

§ 5.27

As has been noticed in § 3.34, Diṅnāga, recognised in general the fallacies based on the dubitability of premises. But in particular he mentioned the fallacies based on dubitability of only the first step of *parārthānumāna*. He does not mention the fallacies based on dubitability of the other premises. Consequently he did not mention the fallacies of instance based on dubitability. Dharmakīrti, however, mentions these fallacies explicitly. To Diṅnāga's list of three fallacies of positive instance,[48] Dharmakīrti adds three fallacies based on dubitability. These are named as *sandigdhasādhyadharma* (where presence of *sādhya* in the positive instance is dubitable), *sandigdhasādhanadharma* (where presence of *hetu* in the positive instance is dubitable) and *sandigdhobhaya* (where both the presence of *hetu* and *sādhya* in the positive instance is dubitable).

Similarly to Diṅnāga's list of three fallacies of negative instance,[49] Dharmakīrti adds three fallacies based on dubitability. These are named as *sandigdhasādhyavyatireka* (where the absence of *sādhya* from the negative instance is dubitable), *sandigdhasādhanavyatireka* (where the absence of *hetu* from the negative instance is dubitable) and *sandigdhobhayavyatireka* (where the absence of *hetu* and *sādhya* from the negative instance is dubitable).

§ 5.28

Dharmakīrti also adds two important fallacies of instance namely *apradarśitānvaya* and *apradarśitavyatireka* to Diṅnāga's list of other fallacies of instance (the fallacies of instance which I called syntactical fallacies, See § 3.40). Though the syntactical fallacies of instance mentioned by Dharmakīrti are prima facie identical with those mentioned by Diṅnāga, these fallacies have special significance in the context of Dharmakīrti's theory of inference. Dharmakīrti's concept of inference is based on his theory of valid inference and in this context, these fallacies can be clearly regarded as formal fallacies. The examples of these fallacies may be stated as follows :

1) *ananvaya*

The mountain is smoky.

Because the mountain is fiery

For exmaple the kitchen is both smoky and fiery.

2) *apradarśitānvaya*

The mountain is fiery

Because the mountain is smoky

For example the kitchen is both smoky and fiery.

3) *viparītānvaya*

The mountain is fiery

Because the mountain is smoky

And whatever is fiery is smoky.

4) *avyatireka*

The mountain is smoky

Because the mountain is fiery

For example the lake is neither fiery nor smoky.

5) *apradarśitavyatireka*

The mountain is fiery

Because the mountain is smoky.

For example the lake is neither smoky nor fiery.

6) *viparītavyatireka*

The mountain is fiery.

Because the mountain is smoky.

And whatever is not smoky is not fiery.

The term *apradarśitānvaya* refers to the positive instance, corresponding to which there is no statement of *anvayavyāpti* [of the form 'whatever has *hetu*- predicate has *sādhya*-predicate' expressed in the argument. Similarly the term *apradarśitavyatireka* refers to the negative instance, corresponding to which there is no statement of *vyatirekavyāpti* [of the form 'whatever does not have *sādhya*-predicate, does not have *hetu*- predicate] expressed in the argument.

Prima facie it appears that *ananvaya* and *apradarśitānvaya* are identical, because the forms are identical. But Dharmakīrti distinguishes between the two. The distinction seems to be that in the case of *ananvaya* the statement of *anvayavyāpti* is not only absent but also untenable. But in the case of *apradarśitānvaya* the statement of *anvayavyāpti* though absent is tenable. Thus *ananvaya* is not only a formal fallacy but also a material fallacy, while *apradarśitānvaya* is merely a formal fallacy.

Dharmakīrti distinguises also between *avyatireka* and *apradarśitavyatireka*. The difference again is that *avyatireka* is both a formal and a material fallacy, whereas *apradarśitavyatireka* is merely a formal fallacy.

Notes

1. See Part I of Chapter III.
2. सर्वोपसंहारेण व्याप्तिप्रदर्शनलक्षणौ साधर्म्यवैधर्म्यप्रयोगौ उक्तौ । *HBT*, Appendix 7, p. 55.
3. See *NB* 3.8.
4. See *NB* 3.9.
5. See *NB* 3.23.
6. See *NB* 3.122.
7. See *NB* 3.10.

7a. We are using 'universal' in one quote because by universal we mean 'in all particular cases'. Buddhist logicians do not recognise universals in Nyāya sense of the term. See Appendix II.

8. See *NB* 3.8.
9. इदमिदानीं सुमहद् व्यसनमायातम्, गुरुभिरभिहितत्वाद् अप्रकाश्यं, न्यायमनुपालयद्भिः असंवरणीयमिति कष्टतरं व्यसनं कथं निर्वोढुं शक्यम् । *HBT*, Appendix 7, p. 55.
10. त्रैरूप्यं पुनर्लिङ्गस्यानुमेये सत्त्वम् एव सपक्ष एव सत्त्वम् असपक्षे चासत्त्वम् एव निश्चितम् । *NB* 2.5, Also see *NBT* under it.
11. See Note 9, Chapter III.
12. कार्यकारणभावाद् वा स्वभावाद् वा नियामकात् ।
    अविनाभावनियमोऽदर्शनान्न न दर्शनात् ।। *PV* 3.31.

12a. For further elaboration see Appendix II.

13. See § 5.7.
14. See *NB* 3.35, 3.36.
15. See *HBT*, Appendix 7, p. 72.
16. Same as Note 14.
17. Same as Note 15.
18. For P_1, see § 3.8. Also see § 3.6.

18a. I.e. $(x)(Hx \supset Sx) \equiv (x)(\sim Sx \supset \sim Hx)$ (The rule of transposition).

19. न पृथक् दृष्टान्तो नाम साधनावयवः कश्चित्, तेनास्य लक्षणं न पृथग् उच्यते गतार्थत्वात् । *NB* 3.121.
20. *NB* 3.24.
21. *NB* 3.25.
22. *NB* 3.8.
23. See *NM* p. 27.
24. See *NM* p. 6, fn. 5.
25. Refer to § 3.7.

26. Same as Note 14.
27. See *ILSM* p. 78.
28. *NB* 3.34.
29. *NB* 3.37-38.

29a. P_3, P_4 etc., as P_1 and P_2, stand for the forms of *parārthānumāna* rather than the examples thereof. For the symbolic expression of these forms, see Appendix III.

30. अत्रापि न कश्चित् क्रमनियमः इष्टार्थसिद्धेरुभयत्र अविशेषात् ।*VN* p. 8.
31. See § 3.13.
32. For Udyotakara, a doctrine of a school of thought has to be justified by some inference and thus the thesis contradicted by *āgama* may be ultimately contradicted by *anumāna*. See *NV*, p. 113.
33. For more details see § 5.10, 7.58.
34. तत्र त्रयाणां रूपाणाम् एकस्यापि रूपस्य अनुक्तौ साधनाभासः, उक्तावपि असिद्धौ सन्देहे वा प्रतिपाद्यप्रतिपादकयोः ।*NB* 3.55-56.
35. See § 3.40.
36. See § 5.8.
37. See § 3.8.
38. तत एवान्वयव्यतिरेकयोः सन्देहाद् अनैकान्तिकः ।*NB* 3.107.
39. Refer to *NB* 3.69-76.
40. The empirical inconsistency is described by Dharmakīrti as similar to one between hot touch and cold touch while the logical contradiction is described by him as similar to one between being and non-being. See *NB* 3.72-75.
41. Refer to *NB* 3.86-92.
42. न हि इष्टोक्तयोः साध्यत्वेन कश्चिद् विशेष इति ।*NB* 3.92.
43. See *HBT*, Appendix 7, p. 68. *Vivakṣitaikasaṅkhyatva* means that property of *hetu*, by which the *hetu* is supposed to be the only one (without any competitor which may lead to the opposite conclusion). Later this CE was called *asatpratipakṣatva* and the corresponding fallacy was called *satpratipakṣa*.
44. *Ibid.*
45. *Ibid.*
46. बाधाविनाभावयोर्विरोधात् ।*Ibid.* Technically *bādhita* may take the form of *sādhāraṇa anaikāntika* or *viruddha*, which occur due to falsity of the *vyāpti*-statement. (See § 5.17).
47. See *HBT*, Appendix 7, p. 71.
48. See § 3.36.
49. See § 3.37.

Chapter VI

Anumāna and Hetvābhāsa According to Jayantabhaṭṭa, Bhāsarvajña and Vācaspatimiśra

Part I
§ 6.1

We have seen in the last three chapters how there was the controversy between earlier Nyāya and Buddhist logicians on the one hand and between earlier Nyāya-Buddhist logicians and Dharmakīrti on the other, on the issue of the nature of *anumāna*. It seems that in Indian logic upto Udyotakara, the practical aspect of reliable reasoning was emphasised, but the logical aspect was ignored, while Dharmakīrti emphasised the logical aspect without ignoring the practical aspect.

It may be seen that practically we are interested in truth or falsity of the conclusions rather than in validity or invalidity of arguments. Most of the times we use analogical reasoning and claim the truth of the conclusion. So far as the conclusion is not falsified or strongly doubted, we believe that the analogical reasoning made by us is a good reasoning, and may be that we regard it as *pramāṇa*. This is the practical attitude towards *anumāna* and Nyāya logicians with Diṅnāga advocated this.

However, in every analogical reasoning there is a risk that the conclusion of the reasoning may be false. This is the case in inductive reasoning as well as in pre-'inductive- generalisation' analogical reasoning.[1] The logicians of analogical reasoning can develop the techniques by which the risk can be minimized. But they cannot reduce it to zero. Accordingly Diṅnāga's efforts in developing a model of strong analogical reasoning tended towards minimization of this risk.

Some logicians like Dharmakīrti hoped that this risk can be nullified if we restrict the scope of *vyāpti*. As we saw in § 5.6 Dharmakīrti believed that we can be sure about the truth of the *vyāpti*-statement if it is based on either *tādātmya* or *tadutpatti*. the inference based on such a *vyāpti*-statement, according to Dharmakīrti is not only valid, but sound. He hoped that a sound inference is not only logically possible but also practically possible. Dharmakīrti thus did not deny the practical aspect but regarded formal aspect as a necessary and major constituent of the practical aspect. Dharmakīrti's efforts for nullifying the risk were, however, unsuccessful because the problem how we can be sure about the truth of the *vyāpti*-statement remained unanswered.[2]

§ 6.2

We find that the nature of *anumāna* became much more complicated in the hands of Nyāya logicians who came after Dharmakīrti. They were not only influenced by the aspect of formal validity as emphasised by Dharmakīrti, but also by the practical aspect expressed by all others. It is evident that the Nyāya logicians were not satisfied with Diṅnāga's formula of three CEs of *hetu*. In § 5.22 we referred to the Nyāya logicians contemporaneous with Dharmakīrti who thought that CEs of *hetu* should be six. Dharmakīrti refuted this programme of Nyāya logicians. But his refutation was significant from the formal point of view. Nyāya logicians did not emphasise the formal aspect as the major part of the practical aspect. Therefore, at the time of Jayantabhaṭṭa (hereafter Jayanta), Nyāya logicians seem to have stuck to the number of CEs of *hetu*, namely five.[3]

§ 6.3

We find in Jayanta's account of *anumāna*, that the formal aspect and the practical aspect are mixed together. We should, for instance, distinguish clearly between Diṅnāga's frame of three CEs which exhibits mainly the practical aspect and Dharmakīrti's frame of three CEs which exhibits mainly the formal aspect.[4] But Jayanta does not seem to have grasped this distinction in a clearcut way. Whenever he tries to refute Buddhists' position, he tackles Dinnāga and Dharmakīrti on par with each other. Dharmakīrti had in fact shown that if we accept the notion of *vyāpti* as universal and necessary relation then more than three CEs of *hetu* are not logically necessary, because *anvaya* and *vyatireka* (the second and the third CE of *hetu*) can be defined in terms of universal *vyāpti*. The Nyāya logicians starting from Jayanta did accept the notion of *vyāpti* as universal and necessary relation, but still accepted five CEs as the necessary CEs of *hetu*.

Vācaspatimiśra (hereafter, Vācaspati) emphasises that *vyāpti* means a natural *(svābhāvika)* relation between *hetu* and *sādhya*.[5] He distinguishes this natural relation from the conditioned relation. The smoke and fire are naturally related in so far as the direction of the relation is from smoke to fire. But when the relation of smoke with fire is to be expressed as 'fire implies smoke'[5a] then the relation should be regarded as conditioned *(aupādhika)*. Fire would imply smoke only under a certain condition (namely that the fire should be created through the wet fuel). The condition is called *upādhi*. Thus Vācaspati explains the notion of *vyāpti* as the unconditioned necessary relation between *hetu* and *sādhya*. But he still advocates the frame of inference consisting of five CEs of *hetu*. Bhāsarvajña also is not an exception to such an attitude.

§ 6.4

The attitude of Jayanta, Bhāsarvjña and Vācaspati, of accepting more CEs than the three stated by Dinnāga, can be justified on some practical ground. As it is true that any reasoning in which *hetu* possesses only three 'Dinnāgian' CEs, may still lead to a false conclusion, the practically useful ways of discovering the fault in such a reasoning can be as follows,

1) The conclusion can be shown to be false by comparing it with our direct experience.

2) Some other doctrine which has been conclusively established, may be contrary to the conclusion.

3) Another reasoning which is at least as much strong as the proposed reasoning, but gives the opposite conclusion may be available.

Out of these three ways, first two ways are the ways of falsifying the conclusion and the third way is the way of raising doubt about the truth of the conclusion. Jayanta and others call the first two ways as *bādha*. (The types of such defective *hetu* in the defective reasoning are called *pratyakṣabādhita* and *āgamabādhita* respectively by Jayanta.) They call the third way as *satpratipakṣatva*. (The type of the defective *hetu* in such defective resoning is called *satpratipakṣa*). In fact all the three ways are concerned with either falsity or dubitability of the conclusion. Thus in addition to the fallacies based on formal invalidity we find here that Nyāya logicians also discuss the fallacies based on falsity or dubitability of the conclusion. Though the fallacies of *bādhita*-type are derived from *pakṣābhāsas* as discussed by Diṅnāga, there is a distinction between the two. *Pakṣābhāsas* are the fallacies of thesis and, in a sense, pre-argumentation fallacies. If a thesis is shown to be fallacious, the further reasonig becomes useless and irrelevant or even impossible. But the fallacies of conclusion are used to show that the reasoning is unsound. In fact the falsity of a conclusion implies that either the premises are false or the argument is invalid. Thus logically speaking, the fallacies of conclusion are reducible either to the fallacies based on falsity/dubitability of the premises, or to the formal fallacies. Nyāya logicians might have been aware of this fact. Still they have separately accounted for the fallacies of conclusion, because the fallacies of conclusion have a separate practical import.

§ 6.5

Thus the five CEs of *hetu* according to Jayanta, Bhāsarvajña and Vācaspati are as follows,

1) *Hetu* is predicated of *pakṣa*-subject. (*pakṣadharmatvam*[6])

2) *Hetu* is predicated of *sapakṣa*-subject. (*sapakṣadharmatvam*)

3) *Hetu* is not predicated of any *vipakṣa*-subject. (*vipakṣād vyāvṛttiḥ*)

4) The conclusion is not falsified (by experience or by some established doctrine) (*abādhitaviṣayatvam*)

5) *Hetu* is not opposed by another *hetu* which demonstrates the negation of the conclusion of the proposed reasoning. (*asatpratipakṣatvam*)

§ 6.6

The question that may be asked here is whether these five CEs serve as the sufficient conditions of sound *hetu*.

Jayanta and others claim that if *hetu* possesses all the three Diṅnāgian CEs, if the conclusion of the respective reasoning is not falsified by any authoritative means and if equally strong counter-reasoning is not available, then *hetu* must be sound.

Let us suppose that an inference is presented, *hetu* is shown to possess all the three Diṅnāgian CEs, the conclusion is not yet falsified by either experience or any established doctrine, and the opponent also is unable to present a counter-argument (may be, because the opponent is not so imaginative or because the data for counter-reasoning is not availale to him). But how can we be hundred per cent sure that the conclusion must be true? The conclusion may still be false.

Hetu's non-possession of the fourth CE is dependent upon the extra-inferencial (mainly perceptive) knowledge of the negation of the conclusion. But as far as the extra- inferencial cognition has not occurred to the concerned debater, nothing can be said about the sure truth or falsehood of the conclusion. So far as the extra-inferencial cognition has not occurred *hetu* can be said to possess the fourth CE, but *hetu's* possession of the fourth CE gives only non-cognition of the falsity of the conclusion, but not the authoritative cognition of the truth of the conclusion.

Hetu's non-possession of the fifth CE is dependent upon the opponent's imaginative power and availability of the data for presenting the counter-argument. Thus if the opponent's imaginative power or the availability of the data is lacking, then it would mean that the proposed *hetu* possesses the fifth CE. But even in that case the conclusion may be false.

It seems that most of the ancient Indian philosophers believed in some kind of *svataḥ prāmāṇya*, though they are actually divided into two groups namely *svataḥ-prāmāṇyavādin* and *parataḥ prāmāṇyavādin*. This is because generally they accept a cognition to be *pramā* (authentic cognition) if it is not falsified (*bādhita*) by some other authentic cognition. Thus here the inference is accepted to be authentic in so far as its conclusion is not falsified by any other authentic means of knowledge. But it is possible that a cognition which is not falsified may get falsified after sometime. Here we should distinguish between falsification and falsifiability. A cognition which is not falsified may be still falsifiable. But to accept a cognition to be true just because it is not falsified, is to accept that the cognition does not need external means of its conclusive proof though it needs so for its disproof. This is the position of one who believes in '*svataḥ prāmāṇya*' (authenticity by itself) and 'parataḥ aprāmānya' (inauthenticity through external evidence). Nonfalsification (*abādhitatva*) as a criterion of authenticity was first emphasised by Mīmāṁsakas, the official advocates of *svataḥ prāmāṇya*. Here Naiyāyikas are seen to follow the same idea of *abādhitaviṣayatva* in the context of *anumāna*.

§ 6.7

Jayanta is made slightly aware of the fact that even five CEs of *hetu* are not the suifficient conditions of sound *hetu*, when he discusses a type of *asiddha*-fallacy, namely *anyathāsiddha*. This fallacy occurs when *hetu* physically resides in *pakṣa* and thus possessess the first CE, and also possesses all the remaining CEs, but does not imply *sādhya*. That is to say, in this case, though it is observed that generally *hetu* exists with *sādhya*, still it is not proved that '*hetu* cannot exist without *sādhya*'. In the discussion on *anyathāsiddha*, some opponent of Jayanta argues that as this fallacy which refers to the case where *hetu* does not imply *sādhya*, is basic to all fallacies, all fallacies are reducible to this fallacy. Thus *anyathāsiddha* may be regarded as the only fallacy. In any case *anyathāsiddha* cannot be reduced to any of the five fallacies. If at all the other five fallacies are to be mentioned along with *anyathāsiddha*, then it should not be included in any of the five but should be regarded as the sixth fallacy.

Jayanta's answer to this criticism is almost orthodox. He suggests that a person like me (i.e. Jayanta) can commit mistakess, but Akṣapāda, the author of *NS*, cannot;[7] and as he has mentioned five and only five fallacies, the so-called sixth fallacy namely *anyathāsiddha* must also be included in one of the five fallacies. Jayanta justifies his position of including *anyathāsiddha* in *asiddha* by saying that though in case of *anyathāsiddha*, hetu resides in *pakṣa*, it does not reside in the same way as a sound *hetu* resides.[8] This justification is obviously farfetched and Jayanta also is aware that it is.[9]

In fac' the fallacy namely *anyathāsiddha* is committed not due to the falsity of the first CE as Udyotakara,[10] Jayanta and others assumed, but it is committed due to the dubitability of the *vyāpti*-statement. The fallacy may be explained as follows. When some *hetu* is used for demonstrating *sādhya*, then if the *hetu* is sound, it must depend upon *sādhya* in the sense that it cannot exist without *sādhya*. In other words, the existence of *hetu* cannot be fully explained or justified unless the existence of *sādhya* with it is accepted. On the contrary if the existence of a particular *hetu* can be justified, without bringing *sādhya* into the picture, then the *hetu* is called, *anyathāsiddha* ('proved otherwise' or 'explained otherwise'). Thus in the case of *anyathāsiddha*, *hetu* does not imply *sādhya* as it does not depend upon *sādhya* and the *vyāpti*-statement thus is dubitable.

However, in Jayanta's frame of inference including five CEs of hetu, we cannot explain this fallacy satisfactorily, because Jayanta's (or for that matter any Nyāya- logician's) frame of five CEs does not account for universal *vyāpti*. We can explain this fallacy in

Dharmakīrti's frame of three CEs because Dharmakīrti's second and third CE are explainable in terms of *vyāpti*. Though Dharmakīrti does not mention this fallacy, he mentions other fallacies where although explicit evidence against *vyāpti* is not found, still *vyāpti* is doubted as false. By mentioning such fallacies Dharmakīrti made a clear distinction between the notion of strong analogical inference and the notion of sound inference. Jayanta, however, while mentioning the fallacy namely *anyathāsiddha* tried to mix these two notions with each other.

Supposing that we add the sixth CE namely that ***hetu*** must not be *anyathāsiddha*, i.e., *hetu* must really depend upon *sādhya*, then only the first CE and the 'sixth CE' will do. This model will resemble Dharmakīrti's model where Dharmakīrti's second and third CE are reduced to our sixth CE.

§ 6.8

However, even though we conceive of sound inference in the manner in which Dharmakīrti did, we have still to face the problem of sure truth of *vyāpti*-statement. This situation expresses a dilemma of inference. This dilemma, which was first presented by Cārvākas, is referred to by Jayanta in his *Nyāyamañjarī*. Cārvākas' argument against the authenticity of *anumāna*, that expresses a dilemma of inference, can be elaborated as follows.[11]

1) If the inference is made from particular to particular (*viśeṣa*), that is to say, if the inference is only analogical inference, then because the universal link (*anugama*) between signifying analogy (*hetu*) and signified analogy(*sādhya*) is not stated in such inference, we cannot accept this kind of inference as an authentic means of knowledge.

(विशेषेऽनुगमाभावात्[12])

2) If, on the other hand, the inference is made from universal (*sāmānya*) to particular then as the conclusion about the particular fact is already contained in the universal *vyāpti*-statement, the

inference is defective due to *siddhasādhana*, that is to say, the inference.is superfluous. (सामान्ये सिद्धसाधनात्[13])

3) It may be argued here that the superfluity of inference is not incompatible with soundness of the inference. On this Cārvākas argue that superfluity of inference would not be a serious fallacy if premises of the inference are proved to be true and the inference is known to be valid. But though the inference from universal to particular is valid, the universal *vyāpti*-statement which is a premise of such inference, cannot be proved to be true. (तद्धतोऽनुपपन्नत्वात्[14]) Therefore we cannot prove any inference to be sound. (अनुमानकथा कुतः ?)[15]

§ 6.9

It can be seen that the first two parts of the above dilemma resemble the two horns of the dilemma of inference presented in Western logic. The dilemma is presented as follows.

If in an inference, the conclusion is not contained in the premise, it cannot be valid; and if the conclusion is not different from the premises, it is useless; but the conclusion cannot be contained in the premises and also possess novelty; hence inferences cannot be both valid and useful.[16]

This dilemma of inference in Western logic is answered by pointing out that though the conclusion of a valid inference is not logically novel, it can be psychologically novel and thus can be felt as useful. In other words, superfluity need not be regarded as a serious logical flaw, because in spite of the superfluity, the conclusion of a valid argument can be informative. Thus the dilemma of inference can be surpassed in the cases where premises are proved to be true and the inference is valid. But the next question that Cārvākas are posing is regarding the truth of the premises.

§ 6.10

It may be noted here that the dilemma of inference that Cārvākas are posing is not a dilemma of inference in general but only of a particular kind of inference, which comes under the category of *anumāna*. *Anumāna* is an inference in which conclusion is a proposition about some unobserved fact, made on the basis of some observed data. We have seen that the form of *anumāna* as accepted by Indian logicians prior to Dharmakīrti, was that of analogical reasoning. So the first part of the dilemma posed by Cārvākas refers to the analogical form of *anumāna*. However, the Nyāya logicians starting from Jayanta are seen to have left aside the earlier form and accepted the valid form of *anumāna* with universal *vyāpti*-statement as a necessary premise. Cārvākas are raising the question regarding the justification of such a universal *vyāpti*-statement, in the third part of their dilemma.

Cārvākas claim that we cannot establish the statement of the form 'Whatever possesses *hetu*, possesses *sādhya*' unless we observe all the past, present and future instanciations of existences and absences of *hetu* and *sāhya*[17]. We have also to observe that every instanciation satisfies the condition that it falls under either of the following three cases:

(1) *Hetu* and *sādhya* co-exist.

(2) *Hetu* and *sādhya* are co-absent.

(3) *Sādhya* co-exists with the absence of *Hetu*.

But to observe all the instanciations of the existences and absenses of *hetu* and *sādhya* is impossible for any human being. Thus the statement of the form 'Whatever possesses *hetu* possesses *sādhya*' cannot be established.

It can be clear from the above formulation of the Cārvākas' argument against the possibility of the knowledge of *vyāpti*, that Cārvākas are thinking in terms of perfect induction. They are arguing that the induction can be authentic only if it is a perfect induction.[18] But we cannot make any perfect induction regarding any 'universal truth'.[19] Therefore the authentic induction of the above form is impossible.

§ 6.11

While answering the dilemma, Jayanta and other Nyāya logicians following Jayanta make two types of arguments. Under the first type they try to argue that whether or not the inference is dilemmatic in nature, we have to make inferences every now and then and many times we also get true conclusions.[20] We cannot remain alive if we do not infer.[21] Thus as it is a fact that we do make good inferences, we have to accept that inference is an authentic means of knowledge.

Bhāsarvajña repeats this argument made by Jayanta and adds that when Cārvākas raise objections against the authenticity of *anumāna* they do it only by believing in *anumāna* as authentic means of knowledge, because no debater can communicate anything to the other debater unless he believes in the inferred truth that the other debater really intends to say so and so. This is therefore a pragmatic contradiction *(svavacanavirodha)* made by Cārvākas.[22]

§ 6.12

This type of argument really does not answer the Cārvāka criticism because Cārvākas may accept without hesitation that we have to make good or bad inferences every now and then and conclusions of many such inferences are later on verified as true. Cārvākas may also say that when they are arguing with Naiyāyikas against the authenticity of *anumāna*, they are believing that possibly they have understood the correct intention of the Naiyāyikas. Thus the inference with 'possible' conclusion should not be confused with the inference with 'certain' assertive conclusion. Naiyāyikas' *anumāna* is an inference with certain assertive conclusion, which according to Cārvākas cannot be proved to be authentic. Thus Cārvākas may practically use the tool of inference for making guess work and still express their disbelief in *anumāna* as authentic means of knowledge.

Though Cārvākas may accept that some inferences with certain, assertive conclusion may be later on verified to give true conclusions, from this, according to Cārvākas, it does not follow that the inferences with true conclusions can be proved to be sound. It may be seen from the Cārvākas' patterns of arguing that Cārvākas must be accepting that formally valid arguments are possible. But the question that Cārvākas are posing is that how we can demonstrate that a particular *anumāna* is not only valid but also sound.[23] It is possible that the conclusion of some valid *anumāna* is later on empirically verified to be true. But the question is whether the inference can demonstrate the truth of its conclusion independently of the empirical verification of the conclusion. The inference cannot do so unless the inference is valid and its premises are demonstrated to be true.

Here the basic question would be how the statement of *vyāpti* which is a necessary premise of any valid *anumāna*, can be demonstrated to be true. Cārvākas do not raise any objection regarding the truth of the other essential premise of *anumāna*, which is called the statement of *pakṣadharmatā* (i.e. the statement which states that *hetu* exists in *pakṣa*), because generally the statement of *pakṣadharmatā* is an observation- statement.

§6.13

The second type of argument that Jayanta and other Nyāya logicians following him, give for answering the dilemma, aims at giving the theory of induction and showing that the authentic inductive reasoning is possible.

Dharmakīrti was perhaps the first Indian logician who was seriously aware of the problem of induction. He tried to answer it by identifying the area in which induction could be done with certainty. The area was that of the relation of effect with the cause *(tadutpatti)* and the relation of having the same reference *(tādātmya)*.[24] But the Nyāya logicians had a problem here; for demonstrating that X is a cause of Y we have to observe many cases where X has existed prior to Y and also see that no case has occurred where Y comes into existence without the prior existence of X. Thus the means of proving the causal relation between *hetu* and *sādhya* is ultimately the observation of co-existence and co-absence. Bhāsarvajña, for instance, objects that if the causal relation is established with the help of the constant conjunction between *hetu* and *sādhya*, then the constant conjunction which indicates that hetu cannot exist without sādhya (avinābhāva) can itself express the universal relation. Thus the causal connection need not be established because *vyāpti* can be grasped without it.[25]

Jayanta saw that even in case of the relation of common reference, *hetu* does not imply *sādhya*[25a] just because *hetu* and *sādhya* have a common reference, but because *hetu* and *sādhya* fulfil the essential condition of *vyāpti*. The essential condition of *vyāpti* according to Jayanta is that '*hetu* exists only with *sādhya* and does not exist without sādhya'.[26] This essential condition of *vyāpti* is known to be fulfilled by a particular *hetu* and *sādhya*, basically by observing co-existence and co-absence. Thus, according to Naiyāyikas, the ideas of tādātmya and *tadutpatti* are superfluous in the broader context of the concept of *vyāpti*, and thus the area of *vyāpti* need not be restricted to the relations of causality and common reference.

§ 6.14

However, Nyāya logicians generally accept that merely the observation of co-existence and co- absence cannot amount to the cognition of vyāpti, but according to them the observation gives rise to a kind of 'mental' perception.[27] Here the 'mental' perception in fact means an intuition. Thus, like some western logicians[28] Indian logicians generally accept that the inductive reasoning is not an inference but an intuition. Nyāya logicians are seen to differ on the question as to what could be the exact nature and status of this intuitive perception in the context of the total Nyāya theory of perception. Still it is always regarded as a perception and is generally distinguished from ordinary perception (*laukika pratyakṣa*).

Of all the Nyāya logicians, Vacaspati firmly rejects the view that *vyāpti* is mentally perceived.[29] Vācaspati thinks that *vyāpti* is known by sensory perception and mind acts in that perception in the same manner as it acts in any other sensory perception. However, the sensory perception of *vyāpti*, according to Vācaspati, is distinct from the other sensory perceptions. The sensory perception of *vyāpti* requires a number of previous perceptions of co-existence and co-absence of *hetu* and *sādhya*, while the other sensory perceptions do not. This view of Vācaspti seems to be rooted in his notion of *vyāpti*. As we saw in § 6.3, Vācaspati emphasised the notion of *vyāpti* as a natural relation. He thought that when the particular objects like smoke and fire are seen through eyes, the 'natural relation' between them also can be perceived through eyes. But here Vācaspati is confusing between the two meanings of *svābhāvika-sambandha*, one as particular, natural, 'perceptible' relation between two perceptible particulars and the other as universal unconditioned (*anaupādhika*) relation between the two objects 'in general'. Vācaspati has to face the objection that *vyāpti* is not a natural relation in the first sense but in the second sense. By sensory perception we cannot even pretend to know all the relata of *vyāpti*, unless it is aided by some kind of intutition. Vācaspati, however, does not discuss this kind of objection.[30] It seems therefore better to say that the cognition of the so-called universal *vyāpti* is an intuition or intuitive perception and not just the sensory perception.

§ 6.15

It seems that the principle behind such intuition of universal law is that 'when a general property of a thing is perceived, all the things, as far as they possess that general property, are mentally perceived'. When I see a tree, I can claim that I have seen 'what after all a thing called tree is'. This claim would be interpreted by Nyāya logicians as a claim that 'I have seen all the trees as far as they are trees'. In the similar fashion Nyāya logicians seem to argue that when one sees smoke and fire related with each other in a certain way, he sees all the cases of that relation between smoke and fire, as far as those are the cases of that particular relation between smoke and fire.

In the above argument, however, the question remains, how we can justify thereby that there cannot be smoke without fire. The above-stated intuition involves an intellectual jump which was neglected by Nyāya logicians. Cārvākas point out the same when they say that even though we observe hundred cases of co-existence, still the possibility that *hetu* may exist without *sādhya* cannot be nullified.[31]

If *vyāpti*-statement is a statement from mental perception and perception is regarded as *pramāṇa*, then just as we have at our disposal, a verification procedure for checking an observation-statement to be true or false, Nyāya logicians should be in a position to provide a similar verification-procedure for the inductive reasoning. But in fact, strictly speaking, we cannot verify any universal statement as we can verify a particular observation-statement. Secondly, just because a particular cognition is an intuition, it does not follow that it must be true. Nyāya logicians, however, fail to provide any method of verification or any criterion by which a perception of *vyāpti* can be judged to be true.

Jayanta expresses his belief that if *vyāpti* is correctly grasped and a good inference is made, then the conclusion cannot be falsified.[32] He adds that if *vyāpti* is not correctly grasped and a bad inference is made, then it is not the fault or weakness of the tool, namely inference, but it is the fault of the user of the tool, namely, the person who infers.[33] This is not an answer, but this is a postponement of the answer, because again the question may be asked "What is the criterion for judging that the inferring person is not at fault?" If the answer is given in terms of the fault in the inference, the circularity is clear.

Vācaspati and Bhāsarvajña also cannot answer the problem satisfactorily. Bhāsarvajña only points out the fact that the 'non-fictitious' cognition of universal laws is a matter of common experience.[34] He does not refer to any procedure by which the cognition of a universal law can be shown to be fictitious or non-fictitious. Vācaspati believes in the common-sense-procedure of verification, that we may try our best to find out the concrete case which would falsify the *vyāpti*-statement, but if such cases are not found, we should believe that the *vyapti*-statement must be true.[35] Thus Vācaspati is again seen to express his belief in a kind of *svataḥprāmāṇya*. But the objection can be raised that the non-availability of falsifying evidence would only increase the probability of the truth of the *vyāpti*-statement, but it could not 'prove' it to be true.

§ 6.16

Jayarāśibhaṭṭa, a heretic philosopher, (though not exactly a Cārvāka-philosopher,[36]) who flourished after Dharmakīrti, but before Jayanta, has referred partially to this dilemma of inference. He also quotes the statements namely *'viśeṣe'nugamābhāvaḥ'* and *'sāmānye siddhasādhyatā.'*[37] He presents the dilemma as follows,

1) If the object which is known by *anumāna* is the same *(samāna)* as one which is known by perception, then *anumāna* is superfluous.[38]

2) If the object known by *anumāna* is different (*viśeṣa*) from one which is known by perception, then the link between the inferred object and the other objects like the observed *hetu*, is not known. Therefore the conclusion will not follow from the reasoning.[39]

It can be easily seen that the first part of this dilemma resembles the second part of the dilemma presented in § 6.8, while the second part of this dilemma resembles the first part of the dilemma presented in § 6.9.

§ 6.17

Bhāsarvajña interprets this dilemma in a different manner. According to him Cārvākas argue that if *sādhya* is of the universal kind, i.e., if it is the universal aspect (for instance, the universal 'fire') of the *sādhya*, then as this universal aspect is already supposed to be proved when the universal *vyāpti* is known, the inference becomes superfluous.[40]

If, on the other hand, *sādhya* is the particular object (for instance, the particular fire on the mountain), then its assertion does not follow from the premises.[41]

The dilemma may be restated in the following way.

1) Suppose, the inference is of the form,

"For every x if x possesses *hetu* then x possesses *sādhya*. y possesses *hetu*/ ∴ y possesses sādhya"

where x and y are variables, then the propositional function namely 'y possesses *sādhya*' is already contained in the premises. Such an inference is superfluous (and also non-informative).

2) If, on the other hand, the inference is of the form,

'For every x if x possesses *hetu* then x possesses *sādhya*. a possesses *hetu*/∴ a possesses *sādhya*'

where x is a variable and a is a constant, then 'a possesses *sādhya*' cannot follow from the premises because it is not contained in the premises. The conclusion is not contained in the premises because the statement about variable (universal) and the statement about constant (particular) are two entirely different types of statements and one cannot be translated in terms of the other. In the first part of the dilemma, both x and y are variables, and thus one can be translated in terms of the other.

It can be seen that this dilemma is not concerned with analogical type of inference, but it is concerned with the so-called valid type of inference. In the second part of this dilemma, the validity of the so-called valid *anumāna* itself is questioned.

§ 6.18

Bhāsarvajña and the later Nyāya logicians consider the problem of superfluity of inference as an important issue. While defending the position that *anumāna* is not superfluous, they seem to assume that the universal statement is of a different type from a conjunction of particular statements.[41a] This assumption seems to reflect the Nyāya insight into the epistemology of *vyāpti*, because when somebody intuits a universal statement he does not necessarily intuit the conjunction of all concerned particular statements, as he does not necessarily intuit all the actual and possible situations.

It is true that Nyāya logicians do not intend to say that the probandum should be a propositional function, but they expect that it should be a proposition. However, they also expect that the predicate of the conclusion (i.e. the *sādhya*) should be of universal kind. Thus Bhāsarvajña argues that the predicate of the conclusion does not express the particular *sādhya*, but the common aspect of *sādhya*. However the common aspect of *sādhya* is not predicated of any veriable (unspecified, universal) subject but of a particular subject namely the *paksa*-subject. Thus the inference is informative.[42]

Thus Bhāsarvajna seems to have answered the problem of superfluity. But the crucial part of the dilemma, again is the dubitability of the truth of the *vyāpti*-statement. And as we have seen in § 6.15, this part of the dilemma does not seem to be answered satisfactorily by the Nyāya logicians.

§ 6.19

A way-out from this problem of inference, if at all it is a way-out, seems to be that we do not deny the risk of inferring, although we cannot help inferring. Thus the practical importance of inference is not denied, but the value of the inference is partly judged by the logical and material strength of the inference and ultimately by the later empirical verification or falsification of the conclusion. Thus the experience should not only be regarded as the basis of all knowledge but also the justificatory principle of all knowledge. It seems that Cārvākas are arguing in the same direction when they regard only *pratyakṣa* as *pramāṇa*. The similar intention of Cārvākas is suggested when some Cārvākas, as referred to by Jayanta,[43] distinguish between two types of inference,the first of which is the inference with empirically verificable conclusion (*utpannapratīti*) and the second is the inference with non- verifiable conclusion (*utpādyapratīti*).[44]

§ 6.20

The first attacks against the authenticity of *anumāna* were made by Vedāntic philosophers. Also Bhartṛhari, who very much believed in the authenticity of Vedas, attacked the authenticity of *anumāna* and discovered the permanent risk of making any *anumāna*.[45] These philosophers attacked *anumāna* with the aim of saving Vedic doctrines from the arguments made by the heretics of that time. However, we find more thorough and systematic attack against the authenticity of *anumāna* appearing first in *Tattvopaplavasiṁha* by Jayarāśibhaṭṭa and then in *Nyāyamañjarī* by Jayanta. *Tattvopaplavasiṁha* in all probability has been written after Dharmakīrti wrote *Pramāṇavārtika*.[46] Thus it can be said that the thorough criticism of *anumāna* as authentic means of knowledge came up when the model of sound reasoning was presented. As there was development in the account of *anumāna*, there was also a similar development in the critique of *anumāna*.

It can be said that by discovering the dilemma of inference Cārvākas at this time expressed important doubts which show that any *anumāna* must be basically fallacious. Here it may be noticed that in Indian logic even an inference which can be reasonably doubted as unsound was declared to be fallacious. That is why, the fallacies where *vyāpti*-statement can be doubted to be true were recognised as genuine fallacies by Dharmakīrti.[47] Cārvākas pointed out that in any inference from universal to particular, we can question the truth of the universal statement and thus can doubt the soundness of the inference; for all such inferences are finally based on induction, which is not a valid type of reasoning.

§ 6.21

We have seen (in §§ 4.3-4) that Udyotakara was the first logician to distinguish clearly between the three types of *hetu* which was based on analogy: purely positive, purely negative and positive-negative. After this distinction was made, it became a controversial issue amongst the Nyāya logicians. These three types of *hetu* can be distinguished as far as the practical aspect of *anumāna* is considered. But when we emphasise the formal aspect of *anumāna*, the distinction between these types is not significant. That is why Dharmakīrti does not refer to this distinction. For Dharmakīrti, every statement of *anvayavyāpti* implies the statement of corresponding *vyatirekavyāpti* and vice versa.[47a] Thus every *hetu* according to him was positive-negative. Naiyāyikas after Dharmakīrti, were influenced by the formal aspect of inference presented by Dharmakīrti, as also by the practical aspect. Thus the importance of the distinction between the three types of *hetu* continued to influence them.

Udyotakara had shown by the example of purely negative *hetu*, that purely negative hetu can be used for proving the existence of soul.[48] But he could not find any suitable example of purely positive *hetu*. He had to borrow the example from Buddhists. But Buddhists' example was based on the Buddhist doctrine that whatever is real is impermanent.

We find that the suitable Nyāya example of purely positive *hetu* was not available even at the time of Jayanta, who saw that the acceptance of purely positive *hetu* may strengthen the position of Buddhists (because through their example of the same, we will have to accept their doctrine of impermanence), rejected the authenticity of purely positive *hetu* altogether. However, Jayanta did not reject purely negative hetu because, as he clearly says, it was a good weapon in the hands of Naiyāyikas for proving the existence of soul etc.[49]

§ 6.22

We can see that as soon as a suitable Nyāya-example of purely positive *hetu* was found by Nyāya logicians, the purely positive *hetu* was accepted as a ginuine *hetu*. Thus Bhāsarvajña had no hesitation in accepting purely positive *hetu* as a genuine *hetu* because the genuine Nyāya-example was available to him. His example runs as follows,

"The objects like *adṛṣṭa* are perceptible by someone.

Because they are knowable.

Whatever is knowable is perceptible by somebody like a myrobalan[50] in our palm."

Here the objects like *adṛṣṭa* are not perceptible by common people, but according to Nyāya, only God can perceive them. Whatever is, is perceptiable by God according to Nyāya. So, according to Nyāya, there is nothing which can be included in *vipakṣa*, because here *vipakṣa* is the class of those objects which are not perceptible by anyone.

Vācaspati, who follows Udyotakara and Bhāsarvajña in accepting the three types of hetu, gives an example of purely positve *hetu* which resembles Bhāsarvajña's example.[51] These examples (namely Bhāsarvajña's and Vācaspati's) of purely positive *hetu* are proposed to prove the existence of God. But atheists would not accept these arguments because they would not accept the universal statement namely whatever is knowable is perceptible by somebody. Therefore in Navya-Nyaya, the example of purely positive *hetu*, which could be acceptable even to atheists, is chosen. We can see that it became a big headache for Navya-Nyāya logicians to define *anumāna* in such a way that it could include the inference from purely positive *hetu*.

The above discussion (in §§ 6.21-22) shows how the extralogical considerations entered and poluted the problem of *kevalānvayi* and *kevalavyatireki* inference.

§ 6.23

The logical problem of *kevalānvayi* and *kevalavyatireki* inference seems to concern itself mainly with the soundness of such inferences. *Kevalavyatireki* inference, for instance, raises at least two issues.

1) Can the *hetu* in such inferences be, properly speaking, purely negative?

2) How can such *hetu* be distinguished from the fallacious hetu called *asādhāraṇa*?

Jayanta seems to be aware of the first issue. He says that the so-called purely negative *hetu* is in fact positive-negative (*anvayavyatirekavān*), but only due to specific characterisation of that *hetu* it becomes *kevalavyatireki*.[52]

Bhāsarvajña rejects this view and says that such *hetu* can appear as positive-negative only if the *vyāpti* is presented in a too general form. But such a too general form of *vyāpti* is not useful in such an inference. Thus the specific characterisation of *hetu* and *sādhya* is to be made. But if it is made, *hetu* in such a case cannot be related with *sādhya* by *anvaya*.[53]

It may be noticed here that in the case of *kevalavyatireki* inference, when we are talking of non-existence of *anvayavyāpti* between specified *hetu* and specified *sādhya*, we are identifying *anvayavyāpti* with *sapakṣasattva*. Thus we are excluding *pakṣa* from the scope of *vyāpti*. But when we consider the form of inference with universal *vyāpti*, no sound *hetu* in such inference can be purely negative, because here the *anvayavyāpti* between *hetu* and *sādhya* exists in *pakṣa*, though does not exist elsewhere. In this sense, purely negative sound inference is impossible.[54]

§ 6.24

The second issue may now be raised here. How can such *hetu* be distinguished from the fallacious *hetu* called *asādhāraṇa*? The simple answer that may be given is that the purely negative *hetu* proves *sādhya* while the fallacious one does not. But here the real question is, how to know whether the given *hetu* proves *sādhya* or not, just by looking at the nature of that *hetu*? This question is difficult because in case of *asādhāraṇa, hetu* resides neither in *sapakṣa* nor in *vipakṣa*. Generally the difference observed between the two is that in the case of the first *sapakṣa* exists, but *hetu* does not exist in it, while in the case of the second, *sapakṣa* is non-existent. (It seems that Bhāsarvajña does not approve of this difference because he conceives of the possibility that in a case of *asādhāraṇa, sapakṣa* does not exist.[55]) Granting this difference, the question remains, what difference does it make to the soundness of such *hetus*? Both these *hetus* do not possess the second CE. Thus it may be argued that both of them are on par. Jayanta refers to this kind of objection but leaves it unanswered.[56]

Bhāsarvajña has included the varieites of *asādhāraṇa* (without calling it so) in *anadhyavasita*. While discussing *anadhyavasita*, Bhāsarvajña distinguishes this fallacious *hetu* from the purely negative *hetu* just by characterising the fallacious *hetu* as that hetu which does not prove *sādhya* (*sādhyāsādhaka*[57]). But he fails to clarify what makes this fallacious *hetu* not prove *sādhya*. Just to say that purely negative *hetu* is not fallacious because it does prove *sādhya* and the fallacious *hetu* called *anadhyavasita* is fallacious because it does not prove *sādhya*, amounts to begging the question.

§ 6.25

A problem may be raised also in the case of purely positive *hetu*. The purely positive *hetu* resides in *sapakṣa*, but does not reside in *vipakṣa* because *vipakṣa* corresponding to such *hetu* does not exist. We have noted in § 3.9 Diṅnāga's suggestion namely if *vipakṣa* does not exist, the third CE is already possessed by the *hetu*. The third CE is possessed by *hetu* either if *vipakṣa* exists, but *hetu* does not exist in it, or if *vipakṣa* does not exist. The suggestion is more or less verbal.

How can one be conclusively sure about the universal relation between *hetu* and *sādhya* just by observing the cases of co-existence or co-obsence, seems to be the basic question in both the cases - purely positive *hetu* and pure negative *hetu*. It seems that in general it was accepted that both co- existence and co-absence should be observed. But Udyotakara thought that *vyāpti* which is known by observing only co-existence or only co-absence can work. This controversy touches the basic problem as to whether *vyāpti* can be conclusively proved at all. The answer seems to be that neither co-existence nor co-absence nor both can give the infallible, riskless knowledge of *vyāpti*. We can talk of less risk or more risk but not of no risk. We can also say that if both co-existence and co-absence are observed there is not much risk. But if only co-existence[58] or only co- absence are observed then there is more risk. Again, out of the two, if only co-absence is observed then there is much more risk because then we are not acquainted with the nature of *sādhya* at all. The inference based on the observation of mere co-existence is in a better position because there we are acquainted with the nature of *sādhya* as well as with the 'positive relation' between *hetu* and *sādhya*.

Part II

§ 6.26

As I have suggested (in § 5.22), the Nyāya logicians prior to Jayanta must have proposed the three additional CEs of *hetu*, and accordingly they must have explained the fallacies called *satpratipakṣa* and *bādhita*. The Nyāya account of *hetvābhāsa* after this stage does not seem to have undergone any crucial change. However, there are some minor differences in the accounts of fallacies proposed by Jayanta, Bhāsarvajña and Vācaspati. In following sections (§§ 6.27-31) we shall be considering these differences by considering the five *hetvābhāsas* one by one.

§ 6.27

Savyabhicāra: Jayanta recognises only one type of this namely *sādhāraṇa*, which is the primary type of of *anaikāntika*.[59] Thus he does not mention the other two types stated by Diṅnāga, namely *asādhāraṇa* and *viruddhāvyabhicāri* as the types of *anaikāntika*. Out of these, *viruddhāvyabhicāri* is replaced by *satpratipakṣa* and is mentioned as a distinct fallacy. Jayanta does not include *asādhāraṇa* in *savyabhicāra*. Neverthless he does not give any alternative position to this fallacy in the total classification of the fallacies. Bhāsarvajña, following Praśastapāda, includes *asādhāraṇa* (and also *anupasaṁbhāri*) in *anadhyavasita* and regards this fallacy as the sixth *hetvābhāsa*. He does not agree with the view that there are only five types of *hetvābhāsa*. He says that though Akṣapāda mentioned only five fallacies of hetu, still he was aware that there were more types, and stated only five because he wanted to indicate the major types of fallacies.[60] Bhāsarvajña mentions eight subtypes of *sādhāraṇa*, all of which are already mentioned by Udyotakara in his long list.

§ 6.28

Viruddha: This fallacy in the hands of these Nyāya logicians does not differ in nature from the same described by Dharmakīrti. As I suggested in § 2.14, by *viruddha* Akṣapāda means an external fallacy.

But Dharmakīrti, Jayanta and the Naiyāyikas after them conceive of this fallacy as an internal fallacy.

Bhāsarvajña refers to eight sybtypes of *viruddha*, all of which are already included in the big classification made by Udyotakara.

§ 6.29

Prakaraṇasama: By introducing the fallacy called *satpratipakṣa*, Jayanta expresses his interest in the practical aspect of *anumāna* and by falsely[61] identifying it with *prakaraṇasama* of Akṣapada he shows how an orthodox Nyāya logician he is. All the Nyāya logicians after Jayanta, follow Jayanta on this issue.

Satpratipakṣa according to Jayanta resembles *viruddhāvyabhicāri* of Diṅnāga. But the two are not identical. The distinction between the two according to Jayanta seems to be that *satpratipakṣa* cannot be included in *anaikāntika*, while Diṅnāga has included *viruddhāvyabhicāri* in *anaikāntika*. Over and above this Jayanta does not make any clear distinction between the two. The difference between the two, however, can be shown as follows.

Though in both the cases we have two *hetus* which prove mutually opposite conclusions, in the case of *viruddhāvyabhicāri*, as it appears from Diṅnāga's version, the same debater states the equally strong but opposite *hetus*, while in the case of *satpratipakṣa*, it is the opponent who states the second (opposite) *hetu* as a response to the first *hetu* provided by the disputant. Apparently it seems that *viruddhāvyabhicāri* is a practical impossibility, because the same debater is not found stating mutually opposing *hetus* in a succession. But *viruddhāvyabhicāri* could be intelligible if we consider a case where the opponent presents the two mutually opposing *hetus* the first stated by the disputant and the second discovered by the opponent, and by juxtaposing them shows how the two mutually opposing but equally strong *hetus* are making the decisive conclusion impossible. But in this case *viruddhāvyabhicāri* will be a type of refutation (because it is totally expressed by the opponent for refuting the disputant's thesis) rather than a type of fallacy. *Satpratipakṣa*, however will be regarded only as a type of fallacy, because the term denotes the fallacious *hetu* stated by the disputant.

Bhāsarvajña on the contrary identifies *viruddhāvyabhicāri* and *satpratipakṣa*. While discussing the fallacy he suggests that in the case of *viruddhāvyabhicāri* one and the same debater does not state the two mutually opposing *hetus*, but the second *hetu* is stated by the opponent while the first is stated by the disputant.[62] Thus Bhāsarvajña interprets *viruddhāvyabhicāri* in such a manner that it would tally with *satpratipakṣa*. If a *hetu* is *viruddhāvyabhicāri* then it means that the *hetu* and its opposite *hetu* invariably *(avyabhicāri)* prove their respective conclusions which are mutually contradictory. Here Bhāsarvajña suggests that both the *hetus* are not *avyabhicāri*; the *hetu* is *avyabhicāri* from the point of view of the particular debater who uses the respective *hetu*.[63] Thus in fact Bhāsarvjña accepts the contention of Dharmakīrti that if a *hetu* has an equally strong opposing *hetu*, both the *hetus* cannot be logically sound, but at least one of the *hetus* must be unsound, due to *savyabhicāra*. Thus satpratipakṣa cannot in fact be regarded as a fallacy distinct from *savyabhicāra*.

However, Bhāsarvajña still regards this as a distinct fallacy from the practical point of veiw. Practically speaking sometimes it may happen that in a certain fallacious argument, possibility of *vyabhicāra* cannot be shown by the regular method, but a good opposing *hetu* may be proposed which functions as a means of demonstrating the possibility of *vyabhicāra*. Thus though both *sandigdha-savyabhicāra* (the possibility of *vyabhicāra*) and *satpratipakṣa* generate a doubt regarding the truth of the *vyāpti*-statement, still the way in which they generate, differs.

Secondly *satpratipakṣa* makes *vyāpti* dubitable indirectly, but makes the conclusion of the first *hetu* dubitable directly. If the conclusion is made doubtful, the purpose of the refutation is served. Thus the opponent need not proceed to say 'not only conclusion is made dubitable, but also the *vyāpti*-statement is made dubitable'. In this way the practical functioning of the satpratipakṣa differs from the practical functioning of *savyabhicāra*. The same thing can be said about the distinction between *bādhita* and *savyabhicāra*.

Though while commenting upon Udyotakara's refutation of *viruddhāvyabhicāri*, Vācaspati, following Jayanta, recognises the distinct fallacy called *satpratipakṣa*, he clearly accepts that ultimately the CE of *hetu* namely *asatpratipakṣatva* is unreal[64] (as a distinct CE).

§ 6.30

Sādhyasama: Udyotakara identified this with *asiddha* though it slightly differs from asiddha as understood by Akṣapāda and Vātsyāyana.[65] Jayanta and all the later Nyāya logicians stick to this identification. Jayanta recognizes many types like *anyatarāsiddha*, *ubhayāsiddha*, *svarūpāsiddha*, *āśrayāsiddha* (and the possible combinations of these), *vyadhikaraṇāsiddha* and *anyathāsiddha*. It can be easily seen that he borrows these types from Udyotakara.

Jayanta also suggests that the so-called *vyadhikaraṇāsiddha* is not always a fallacy. Sometimes the substratum of *sādhya* may be different, but still there may be *vyāpti* between *hetu* and *sādhya*. This is an important suggestion because this rules out that the substratum of *hetu* should be the same as the substratum of *sādhya*. Thus, for instance, the inference of the form

"For every x if x possesses *hetu* then b possesses *sādhya*. c possesses *hetu* / Therefore b possesses *sādhya*" where x is a variable and b and c are constants referring to two different substrata, can be a sound inference. Jayanta's and others' recognition of *anyathāsiddha* as a subtype of *asiddha* is, as I showed in § 6.7, questionable.

Bhāsarvjña describes twelve types of *asiddha*, all of which are again borrowed from Udyotakara's classification.

§ 6.31

Kālātīta: Jayanta and other Nyāya logicians after Jayanta have identified *kālātīta* with *bādhita* by trying to show that etymologically *kālātīta* means *bādhita*. But these efforts are clearly artificial. As we have seen, *kālātīta* is based on the confusion about the time of *hetu's*

occuring/appearing, while *bādhita* is based on direct falsification of the conclusion. Unlike *kalatita, bādhita* is in fact a modification of the *pakṣābhāsas* described by Dinnāga. Udyotakara and the Naiyāyikas following him (including Jayanta and Vācaspati), however, refused to accept that there are any fallacies of thesis (i.e. *pakṣābhasas*). Udyotakara argued that the so-called fallacies of *pakṣa* (or *pratijñā*) are really the faults of arguer[66] and not the faults in arguments. Bhāsarvajña, however, recognises the fallacies of *pakṣa* or *pratijñā* and identifies them with the fallacies of *hetu*. He says that even the fallacies of *hetu* are ultimately the faults of the arguer.[67]

Logically speaking, *bādhita* cannot be regarded as a fallacy distinct from *savyabhicāra*. But practical functioning of *bādhita* is different from that of *savyabhicāra*. When *savyabhicāra* is discovered in an argument, it directly falsifies the *vyāpti*-statement which is a premise of the argument, and indirectly creates doubt regarding the truth of its conclusion. *Bādhita* on the contrary directly falsifies the conclusion and indirectly falsifies the *vyāpti*-statement.

Regarding the classification of *badhita* Jayanta differs from Dinnāga and Dharmakīrti both of whom accept *anumānaviruddha* as a kind of *pakṣābhaṣa*. Jayanta does not accept *anumānabādhita* as a kind *bādhita* because according to Jayanta an *anumana* cannot refute another *anumāna*. Bhasarvajña however follows Dinnāga and Dharmakīrti in this respect and accepts *anumānaviruddha* as a kind of *bādhita*.

Bhāsarvajña also notes that *anumānaviruddha* is not the same as *satpratipakṣa*. When the fallacy called *anumānaviruddha* is discovered in an argument, the conclusion of that argument is falsified by the stronger (and probably sound) counter- argument, while *satpratipakṣa* makes the conclusion only doubtful.[68]

Jayanta recognises only two types of *bādhita* namely *pratyakṣabādhita* and *āgamabādhita*. Bhāsarvajña recognises the following six types.

1. *pratyakṣaviruddha* (Conclusion falsified by perception)

2. *anumānaviruddha* (Conclusion falsified by inference)

3. *āgamaviruddha* (Conclusion falsified by an authoritative statement)

4. *pratyakṣaikadeśaviruddha* (Conclusion partially falsified by perception)

5. *anumānaikadeśaviruddha* (Conclusion partially falsified by inference)

6. *āgamaikadeśaviruddha* (Conclusion partially falsified by an authoritative statement).

According to Bhāsarvajña all the other types of *pakṣābhāsa* which Diṅnāga recognises can be included in these six.

§ 6.32

As Akṣapāda has not noticed the notion of fallacious instance, it is a problem for an orthodox Nyāya logician to account for the notion. Jayanta is seen to be more othodox than Bhāsarvajña. Jayanta makes a remark that the fallacies of instance (*dṛṣṭāntābhāsas*) are included in the fallacies of *hetu* and that is why Akṣāpada does not mention them. But Jayanta does not show how these fallacies are so included.

Bhāsarvajña, a more liberal logician of Nyāya, mentions all the fallacies of instance accepted by Dharmakīrti. He, following Praśastapāda[69] also adds a fallacy called *āśrayahīna* in which the subject of instance is non-existent. This fallacy distinguishes Bhāsarvajña's position from that of Dharmakīrti who would accept even a non-existent object as a proper instance. It is sufficient for Dharmakīrti if the instance serves the explanatory function properly. But Bhāsarvajña insists that the instance should also be existent. Bhāsarvajña's fallacies of instance which come under *sandigdhāśraya* also are worth noting in this context.

While discussing the fallacies of instance, Bhāsarvajña distinguishes between syntactical fallacies and semantical fallacies, which he calls respectively as *vacanadoṣa* (or *śabdadoṣa*) and *arthadoṣa*. The fallacies of instance namely *ananvaya* and *avyatireka* (which are the two types of *viparītavyāptyabhidhāna* as Bhāsarvajña calls them) are grouped under syntactical fallacies and the remaining are grouped under semantical fallacies. As we have noticed in § 5.28, the syntactical fallacies (in which *nyūna* also can be included) are formal fallacies; while the remaining fallacies (may be, the fallacies of instance or *hetu* or thesis) are semantical fallacies.

Notes

1. See § 2.3
2. See § 5.5, Also see §§ 6.13-16.
3. This is interesting to see why Nyāya logicians must have stuck to the number of CEs viz. five. One of the reasons seems to be that Naiyāyikas being thoroughly orthodox scholars tried always to show that whatever developed logic they had was already expounded by Akṣapāda in *NS*. Akṣapāda accepted five *hetvābhāsas*, therefore CEs corresponding to them also must be five. The later Naiyāyikas even misinterpreted Akṣapāda's account of *hetvābhāsas* for showing that the current account tallies with it. For instance, for Jayanta's misinterpretation of *viruddha*, See § 2.13.
4. For the distinction between Diṅnāga's and Dharmakīrti's accounts of three CEs, See § 5.4.
5. स्वाभाविकत्वं सम्बन्धस्य निश्चिनुमः ।; अतो नियतः सम्बन्धोऽनुमानाङ्गम् । See *NVT* Pt.I, p. 165.

5a. See Note 25a of this chapter.

6. Here it is not forgotten that Naiyāyikas use the term *pakṣa* for Diṅnāga's *pakṣa*-subject, *sādhya* for Diṅnāga's *pakṣa*- predicate, *sapakṣa* for Diṅnāga's *sapakṣa*-subject and *vipakṣa* for Diṅnāga's *vipakṣa*- subject. However, as the difference is only terminological I have continued with the terminology of *pakṣa*-subject etc. which I used in Chapter III and onwards. This is done with the purpose of avoiding complications.
7. तर्हि षष्ठ एवायं हेत्वाभासः ..... अथवा न सूत्रकारनियमोल्लङ्घनं श्रेयः । प्रमाद्यन् खलु परिमितदर्शी मादृशोऽन्यथा ब्रूयात् न न्यायमहोदधेः पारदृश्वा भगवान् अक्षपादः । *NMr*, Pt. II, p. 166.
8. यथा सिद्धः सिद्धो भवति तथा न सिद्धः अन्यथा तु सिद्धः । *NMr*, Pt. II, p. 166.
9. उत्तानवृत्तिमात्रलम्भेऽपि, *Ibid.*
10. Udyotakara includes *anyathāsiddha* in *asiddha*. (See *NV* pp. 174-5)
11. The delimma presented by me is the explanation of the following verse quoted by Jayanta,

विशेषेऽनुगमाभावात् सामान्ये सिद्धसाधनात् ।
तद्वतोऽनुपपन्नत्वाद् अनुमानकथा कुतः ।।
NMr. Pt. I, p. 108.

12. *Ibid.*

13. *Ibid.*

14. *Ibid.* Here *tadvat* means *sāmānyavat* i.e. a universal truth.

15. See *Ibid.*

16. See *ILSM*, p. 173.

17. भवन्नप्यविनाभावः परिच्छेत्तुं न शक्यते ।
जगत्त्रयगताशेषपदार्थालोचनाद् विना ।। and elsewhere in *NMr*, Pt. I, pp. 108-9.

18. Perfect induction means establishing a universal proposition by an exhaustive enumeration of all the instances which are subsumable under it. See *ILSM* p. 275.

19. I have dressed universal truth in inverted commas to distinguish it from the truths conveyed by the universal statements like 'All the tables in this room are black'. Such a universal statement can be known by perfect induction. (Cārvākas may argue here also that this universal statement cannot be demontrated unless all the non-black things in the word are observed not to be the tables in this room'. I suppose that this is an over-expectation). But the statement like 'All the tables (in the world) are black' is a proper type of inductive general statement (though may be false), that Cārvākas are conceiving of. This kind of induction of a 'universal truth' cannot be practised as a perfect induction.

20. अबलाबालगोपालहालिकप्रमुखा अपि ।
बुध्यन्ते नियतादर्थादर्थान्तरमसंशयम् ।। *NMr*, Pt. I, p. 110

21. अनुमानापलापे तु प्रत्यक्षादपि दुर्लभा ।
लोकयात्रेति लोकाः स्युः लिखिता इव निश्चलाः ।।*NMr*, Pt. I, p. 110. Also, अथ धूमानुमानस्यापि तद् दूषणमेव इष्यते तदा सर्वलोकव्यवहारविरोधः । *NBu* p. 223.

22. See *NBu* pp. 223-4.

23. For the distinction between 'valid' and 'sound' see § 5.1.

24. See § 5.6.

25. अथ अविनाभावित्वेन कार्यत्वं, तदसिद्धौ अपि अविनाभावित्वस्य सिद्धेः । *NBu*, p. 291.

25a. The language that *hetu* implies *sādhya* may be felt odd because implication is a relation between statements and not between things or properties. However here and at many other places *hetu* and *sādhya* are used by me as if they are elliptical statements.

26. तस्मिन् सत्येव भवनं न विना भवनं ततः ।
अयमेवाविनाभावो नियमः सहचारिता ।। *NMr*. Pt. I, p. 110.

27. मनसा नियमज्ञानसिद्धेः । *Ibid*, p. 112. Also, चक्षुरादिजनिताग्निविशिष्ट-धूमज्ञानसहितस्य मनसः सर्वोपसंहारेण व्याप्तिग्राहकत्वम् ।*NBu*, p. 220.

28. For instance, W.E. Johnson, See *ILSM*, p. 275.

29. तदिदमवधारणं न मानसम् । अनपेक्षस्य मनसः बाह्ये प्रवृत्तै अन्धबधिराद्यभावप्रसङ्गात् । *NVT*, Pt. I, p. 166.

30. Bhāsarvajña refers to this kind of objection when he says, यद्येवं, ययोरविनाभावो गृहीतः तयोरेव गम्यगमकभावोऽस्तु न अपूर्वाग्निधूमयोः तयोरदृष्टसम्बन्धत्वात् । *NBu*, p. 219.

31. शतकृत्वोऽपि तद्दृष्टौ व्यभिचारस्य सम्भवात् ।*NMr*. Pt. I, p. 112.

32. सम्यगवधृतायां व्याप्तौ विप्लवाभावात् । *Ibid*. p. 112.

33. प्रमातुरेव तत्रापराधो नानुमानस्येति । *Ibid*, p. 112.

34. यदपि अविनाभावसम्बन्धस्य ग्रहीतुम् अशक्यत्वाद् इत्यादि उक्तं तदपि अयुक्तम् । यतोऽस्ति तावदियं लौकिकपरीक्षकाणामति प्रतीतिः अभ्रान्ता निश्चयात्मिका । etc. *NBu*, p. 216.

35. तस्माद् उपाधिं प्रयत्नेन अन्विष्यन्तः अनुपलभमानाः नास्ति इत्यवगम्य स्वाभाविकत्वं सम्बन्धस्य निश्चिनुमः ।*NVT*, Pt. I, p. 165.

36. See my article entitled 'The Philosophical Position of Jayarāśibhaṭṭa' published in IPQ, Vol. V, No. III.

37. See *TUS* pp. 72, 73, 82, 89. Compare these statements with the first line of the verse quoted in Note 11.

38. न च प्रत्यक्षेण अवगम्यमाने अनुमानं न्याय्यम् .... सामान्यं समानता प्रत्यक्षानुमानविषयत्वेन अर्थस्य ।*TUS*, p. 72.

39. विशेषो हि अनुमानैकग्राह्योऽर्थः तेन सह सम्बन्धो नावगतः तदनवगतौ न लिङ्गात् लिङ्गिगतिः ।*TUS*, p. 73.

40. सामान्यस्य अग्निमत्त्वानित्यत्वादिलक्षणस्य क्वचित् सिद्धत्वात् व्यर्थं तत्साधनम् ।*NBu*, p. 211.

41. देशादिविशिष्टस्य अनन्वयादेव असिद्धिः । नहि पर्वतस्थेन अग्निना शब्दगतेन च अनित्यत्वेन अन्वयः कश्चिद् अस्ति ।*NBu*, p. 211.

41a. I.e. (x)(Hx ⊃ Sx) cannot be translated as the conjunction of (Ha ⊃ Sa), (Hb ⊃ Sb), (Hc ⊃ Sc) etc. where a, b, c.... are constants.

42 See the section entitled *'anugamānupapattinirākaraṇam'* in *NBu*, pp. 215-16. Secondly, Bhāsarvajña would argue that though (x) (Hx ⊃ Sx) cannot be 'translated' as (Ha ⊃ Sa). (Hb ⊃ Sb) ...., still we can argue that in a non-empty universe (x) (Hx ⊃ Sx) necessarily implies (Ha ⊃ Sa).(Hb ⊃ Sb).....

43. सुशिक्षिततराः प्राहुः द्विविधमनुमानं किञ्चिदुत्पन्नप्रतीति किञ्चिदुत्पाद्यप्रतीति । *NMr*, Pt. I, p. 113.

44. This distinction is similar to the distinction between *dṛṣṭa* and *sāmānyato dṛṣṭa* inferences, made by Mīmāṁsakas, Vaiśeṣikas etc., who accepted both the types as *pramāṇa*.

45. See *VPB*, Verses 30-42.

46. Many verses from *PV* are quoted by Jayarāśibhaṭṭa. See Appendix 3 of *TUS*. Also see the introduction by SANGHAVI and PARIKH (See *TUS*, p. ix.)

47. See §§ 5.18-19.

47a. It follows from the rule of transposition.

48. Udyotakara's example of this type runs, 'Any living body is not soulless because otherwise it would not have *prāṇa* etc.' See *NV*, p. 123.

49. अथवा यथाकथञ्चिद् वा व्याख्यास्यामः केवलव्यतिरेकिनं त्वीदृशम् आत्मादिप्रसाधने परमम् अस्त्रम् उपेक्षितुं न शक्नुमः इति अयथाभाव्यमपि व्याख्यानं श्रेयः ।*NMr*, Pt. II, p. 138.

50. For this example of purely positive *hetu*, See *NBu*, p. 302.

51. For Vācaspati's example see *NVT*, Pt. I, p. 179.

52. क्वचित् साध्यविशेषे विशेषणविशेषवशात् केवलव्यतिरेकिताम् अवलम्बते ।*NMr*, Pt. II, p. 137.

53. विशिष्टसाध्येन अन्वयाभावात् ।*NBu*, p. 306.

54. MULLATTI does recognise this peculiarity of the so-called purely negative *hetu*. See *NNTI*, Section §4.13.

55. ननु अविद्यमानसपक्षो विद्यमानविपक्षः पक्षव्यापकः पक्षैकदेशवृत्तिश्च अनध्यवसितभेदौ कस्मान्नोदाहृतौ ? ..... अनध्यवसितभेदौ एतौ भवताम् ।*NBu*, p. 316.

56. See *NMr*, Pt. II, p. 137.

57. साध्यासाधकः पक्ष एव वर्तमानः अनध्यवसितः । *NBu*, p. 309.

58. An inference based on only co-existence is regarded as fallacious due to *Sandigdhavyatireka* by Dharmakīrti.

59. *Sādhāraṇa* is the primary type of *anaikāntika*. For instance, Akṣapāda and Vātsyāyana mention only this type, and not *asādhāraṇa* etc.

60. तस्मान्न पञ्चत्वावधारणार्थं सव्यभिचारविरुद्धेत्यादि सूत्रं किन्तु निदर्शनार्थम् ।*NBu*, p. 309.

61. See § 2.17.

62. प्रतिवादी द्वितीयं प्रयुङ्क्ते ।*NBu*, p. 320.

63. अव्यभिचारित्वं चोभयोः प्रतिपत्त्रभिप्रायेण न परमार्थतः ।*NBu*, p. 320.

64. अन्ततोऽसत्प्रतिपक्षस्यैव रूपस्याभावात् इत्यर्थः ।*NVT*, Pt. I, p. 335.

65. For the difference between *sādhyasama* and *asiddha* see § 2.19.

66. सत्यममी नार्थदोषा न वचनदोषाः किन्तु पुरुषदोषाः ।*NV*, p. 114.

67. ते खलु प्रत्यक्षादिविरोधाः प्रतिज्ञाद्वारेण उद्भावयितुं विवक्षिताः तदा हेतुदोषाः इति ।*NBu*, p. 309. Also, सर्वं साधनं स्वविषये समर्थं तद् यदि अविषये पुरुषेण प्रयुक्तम् अर्थसिद्धिं न करोति । ततः पुरुष एव अपराध्यते .... न तु अचेतनस्य साधनस्य कश्चिदुपालम्भोऽस्ति ।*NBu*, pp. 318-9.

68. तस्मादनुमानं बाधकत्वेनोक्तं यदा बाध्यबाधकबुद्धिं जनयति तदा अनुमानविरोध उच्यते । यदा पुनस्तुल्यलक्षणत्वेन उपलभ्यमानं संशयं जनयति तदा विरुद्धाव्यभिचारी एवं च कार्यभेदाद् भेदः इति /*NBu*, p. 212.

69. Praśastapāda talks of *āśrayāsiddha* as a type of fallacy of instance. See *PB*, p. 123.

Chapter VII

Chala, Jāti, Nigrahasthāna, Prasaṅga, Nayābhāsa and Fallacies of Definition

§ 7.0

Naiyāyikas include all types of defective reasoning in *hetvābhāsa* whereas Buddhists include them in *sādhanābhāsa*. But some terms which refer to defective arguments are not seriously taken into account at all either under *hetvābhāsa* or *sādhanābhāsa*. Some such terms are *chala, jāti* and *nigrahasthāna*. These terms are generally not regarded as referring to fallacies (i.e. either *hetvābhāsas* or *sādhanābhāsas*) although they are regarded as some bad reasoning which a good debater should normally avoid. But actually many of them are fallacies. The difference between *chala* and *jāti* on the one hand and *hetvābhāsa* on the other seems to be that *chala* and *jāti* are intentionally committed for confusing and thereby defeating the other debater whereas *hetvābhāsas* may not be committed intentionally at all. In other words one can say that when *hetvābhāsas* are intentionally committed for confusing and thereby defeating the other debater, they fall under *chala* or *jāti*. The motive of victory and defeat is a necessary constituent of hostile debate. Thus the idea of hostile debate is at the background of the concepts of *chala* and *jāti*. Of course *chala* and *jāti* again differ from *nigrahasthāna*, because a *nigrahasthāna* also is not intentionally committed; one does not generally have the intention of one's own defeat. Still the idea of *nigrahasthāna* joins *chala* and *jāti* because along with the two, it does have the background of hostile debate; the idea of *nigrahasthāna* is inseparable from the ideas of victory and defeat in the debate. I feel that *chala, jāti* and some of the *nigrahasthānas* should be called fallacies of controversy and may be distinguished from the other fallacies which are declared to fall under *hetvābhāsa* or *sādhanābhāsa*. Of course when we stick to the basic concept of *sādhanābhāsa*, we have to regard these fallacies of controversy as subtypes of *sādhanābhāsa*. We shall discuss this point in the next chapter.

Similarly *prasaṅga*-fallacies and the fallacies of definition are not explicitly regarded as *hetvābhāsas*/*sādhanābhāsas* as they possess some peculiar characteristics not shared by the other fallacies. However, these too are fallacies and the basic concept of *sādhanābhāsa* can be applied to them as well.

Nayābhāsas again are the fallacies which are historically neglected and kept away from the main stream of the theory of fallacies. *Nayābhāsas* of the Jainas are not mentioned by Naiyāyikas or Buddhists. *Nayābhāsas* resemble *chala* in so far as they are based on ambiguity of language. But *nayābhāsas* do not presuppose the background of hostile debate. Secondly *nayābhāsas* are not intentionally committed. Of course, as we shall see in the next chapter, the basic concept of *sādhanābhāsa* is applicable to *nayābhāsa* as well.

Thus in the following six parts of this chapter we shall discuss the following concepts concerned with fallacies, as they are given in ancient Indian logic : (1) *Chala* (2) *Jāti* (3) *Nigrahasthāna* (4) *Prasaṅga* (5) *Nayābhāsa* (6) Fallacies of definition.

Part I : Chala
§ 7.1

Caraka had referred to *chala* (quibbling) before Akṣapāda.[1] Caraka, however, refers to only two types of *chala* namely *vākchala* and *sāmānyacchala*. Akṣapāda adds *upacāracchala* to the list and tries to show that this third type is not reducible to any of the other two types.[2]

Any *chala* is based on the deliberate misinterpretation of the sentence uttered by the other debater. The differences amongst the types of *chala* are due to the differences in the ways of misinterpretations followed in them. In the following sections §§ 7.2-4 we shall consider three dialogues which are respective examples of the three types of *chala*.

§ 7.2

Vākchala :

Disputant[2a] : Mr. X is *navakambala.*

(Intended meaning : The blanket possessed by Mr. X is new)

Opponent : You said 'Mr. X is *navakambala*'.

(The misinterpretation : Mr. X possesses nine blankets). But Mr. X possesses only one blanket. Therefore your statement is false.

In the above dialogue the opponent intentionally confuses one literal meaning for the other. Here the word *navakambala* can have any of the two literal meanings, 'one who possesses a new blanket' or 'one who possesses nine blankets'.

§ 7.3

Sāmānyacchala :

Disputant : If Mr. X is a good Brahmin, it is quite likely that he bears good knowledge and moral conduct.

(Intended meaning : quite likely = most probable)

Opponent : You said that it is quite likely.

(The misinterpretation : it is quite likely = is certainly follows). Thus according to you even if a Brahmin is crazy, being a Brahmin he has good knowledge and moral conduct. But this is clearly false, therefore you statement is false.

In the above example the expression 'it is quite likely' (*sambhavati* in Sanksrit) does not have two literal meanings. But the opponent deliberately confuses for the literal meaning of it, the meaning which resembles the literal meaning. What the disputant means is the 'general' rule but the opponent intentionally misunderstands 'general' rule for universal law.

§ 7.4

Upacāracchala :

Disputant : See, the chairs are crying.

(Intended meaning - the persons sitting in the chairs are crying).

Opponent : You said that the chairs are crying. (The misinterpretation : The unconscious things namely chairs are crying). But the chairs cannot cry. Therefore you statement is false.

In this example also the word 'chairs' does not have two literal meanings. Nor is there any significant similarity between the 'intended meaning' (of the disputant) and the misinterpretation. But here the intended meaning is the 'implied' meaning ('*lakṣyārtha*') of what the disputant said. Opponent however takes his statement literally.

§ 7.5

All these types of *chala* are the types of refutation. Though they masquerade as good refutations, in fact they are fallacious refutations. In this sense they can be classified under *jāti* (the fallacious refutation : *dūṣaṇābhāsa*). Though many exmaples of *jāti* are the analogical arguments made for refuting the other debater's position, *jāti* need not be always an analogical argument. *Jāti* literally means genus or similarity. This suggests that *jāti*-argument is 'similar' to a good refutation (though actually not a good refutation). The Buddhist term *dūṣaṇābhāsa* clearly conveys this meaning. *Chala* can thus be included in *jāti*. We shall also find that Vātsyāyana's example of the *jāti* called *nityasama* is nothing but a quibbling on the word *nitya*.[3] This suggests that the distinction between *chala* and *jāti* was not clearly made by Akṣapāda and Vātsyāyana. We can, however, say that *jāti* refers to the class of fallacious refutations in general, of which *chala* is a subclass. If a fallacious refutation is caused by deliberate misinterpretation of what the other debater has said, then it is *chala*, otherwise it is merely *jāti*.

Part II : *Jāti*
§ 7.6

As stated in § 1.3, Caraka had referred to the notion of *uttara* which resembles the notion of *jāti*. However, Caraka does not discuss various types of *jāti*. But Akṣapāda refers to many kinds of *jātis* in *NS*, 5.1. Out of the twenty four types of *jāti* referred to by Akṣapāda, many types are not general forms of *jāti* but are merely particular examples of *jāti*. Many general forms of *jāti* stated by Akṣapāda have a close resemblance with each other and the difference between them is slight or superficial. As shown by Udyotakara and others, many times making a *jāti*-argument amounts to 'wrongly accusing the other debater of committing a *hetvābhāsa*'.[4] However, all the examples of *jāti*-arguments are not strictly found to be fallacious refutations or wrong accusations of fallacy. Sometimes it is found that *jāti*-argument exhibits a genuine refutation but the answer given by denying the refutation is unsatisfactory. (See for instance § 7.16). It is possible that under the heads *jāti*[5] and *chala* Akṣapāda was mainly interested in collecting the forms and examples of sophistical refutations (though not always fallacious refutations).

In the following sections (§ 7.7-30) we shall consider the types of *jāti* one by one. While giving the account I shall follow Vātsyāyana and then (in § 7.31) make a few remarks regarding the later refinement of the account of *jāti*. While considering every *jāti*, I shall first present dialogue or dialogue-form in which an opponent makes a *jāti*-argument for refuting the disputant's argument, and the *jāti*-argument is again answered by the disputant. After presenting the dialogue or the dialogue-form I will make, when necessary, some comments on it.

In dialogue forms we use following symbols for convenience. Generally we use -

a for *pakṣa* - subject

H for *hetu*

S for *sādhya*

~S for the absence of *sādhya*

d for subject of the positive instance.

d' for subject of the negative instance.

e for subject of the counter-instance.

F for the counter-*hetu*.

§ 7.7

Sādharmyasama :

Disputant : a has S, because things having H have S. For instance d has both H and S. Similarly a has H, thus it has S.

Opponent : a does not have S because things having F don't have S. For instance e has both F and ~ S. Similarly a has F. Thus it does not have S.

Disputant : Having F is not the sufficient reason for not having S. F is only the positive analogy between e and a.

Comments : This type of *jāti* is a 'wrong accusation of the fallacy namely *prakaraṇasama* (*satpratipakṣa*)' as described by Vācaspati.[6] Udyotakara calls this as a wrong claim of the fallacy namely *anaikāntika* (*anaikāntikadeśanābhāsa*[7]). It was natural because at the time of Udyotakara the fallacy '*satpratipakṣa*' was not known by its name. *Viruddhāvyabhicāri* which resembled *satpratipakṣa* was known to Udyotakara through Diṅnaga. But Diṅnaga had included it in *anaikāntika*.

BARLINGAY states that this *jāti* is the same as *satpratipakṣa*.[8] But the fact seems to be that making *jāti*-argument of this type does not necessarily mean 'committing the fallacy *satpratipakṣa*', but it means 'making a false claim of the fallacy *satpratipakṣa*'. This *jāti* itself refers to a fallacious argument because it is based on mere analogy which is not strong enough. Disputant's answer suggests the same thing.

It should be noted here that in all the examples of *jātis* being presented here, even the arguments made by the disputants, which are attempted for refutation by the opponents, are analogical arguments. This is because the nature of inference accepted by Vātsyāyana was that of analogical reasoning. However when the analogical reasoning was supposed to be strong enough to rely upon, it was regarded as *pramāṇa* and when it was a weaker analogical reasoning attempted for refutation, then it was regarded as *jāti*.

§ 7.8

Vaidharmyasama :

Disputant : a has S because the things not hainvg S do not have H. For instance d' has neither S nor H. But a has H, therefore a has S.

Opponent : a does not have S because the things having S do not have F. For instance e' has S but not F. But a has F, thus a does not have S.

Disputant : Having F is not the sufficient reason for not having S. F is only the negative analogy between e' and a. Thus your refutation is not sound.

Comments : This *jāti* is similar to the first one as far as both mean the wrong claims of *satpratipakṣa*. The only difference is that in case of *sādharmyasama* the analogy used by the opponent is positive, while in case of *vaidharmyasama* the analogy used by the opponent is negative. It is possible to conceive of another example of *sādharmyasama* in which the disputant argues from negative analogy and is responded by the opponent by using positive analogy. Similarly it is possible to conceive of another example of *vaidharmyasama* in which the disputant argues from positive analogy and is responded to by the opponent by using negative analogy.

§ 7.9

Utkarṣasama :

Disputant : a has S because things having H have S. For instance d has both H and S. Similarly a has H. Thus it has S.

Opponent : If according to you a is like d, then as d has F, similarly a should have F. But a does not have F. Therefore your assumption that a is like d must be false.

Disputant : Having F is not the sufficient reason for not having S. F is merely a negative analogy between a and d. You have not disproved that a has S.

Comment : The *jāti*-argument of this type is a *prasaṅga*-agrument. A debater is supposed to make a *prasaṅga*-argument, for instance when he accepts (for a while) the position of the other debater partially, and shows how it leads to some unacceptable consequence. We find that this was a popular method of refutation which may be called an indirect method of refutation. We shall find here that many forms of *jāti*-argument are the forms of *prasaṅga*-argument.

§ 7.10

Apakarṣasama :

Disputant : a has S because things having H have S. For instance d has both H and S. Similarly a has H. Thus it has S.

Opponent : If according to you a is like d, then as d does not have F, similarly a should not have F. But a has F. Thus your assumption that a is like d must be false.

Disputant : Not having F is not the sufficient reason for not having S. F is merely a negative analogy between d and a. You have not disproved S.

Comment : The *jāti*-argument of this type also is clearly a *prasaṅga*-argument.

§ 7.11

Varṇyasama :

Disputant : a has S because things having H have S. For instance d has both H and S. Similarly a has H. Thus it has S.

Opponent : Before your argument, a is not proved to have S. And as according to you d is like a, d also must not be proved to have S. Thus your assumption that d has both H and S must be wrong.

Disputant : According to me d is not like a in all respects. Though d and a are similar because both are proved to have H, still they are dissimilar because d is already proved to have S, when a is not already proved to have S, but is being proved to have S.

Comments : In § 1.6 we saw that Caraka introduced a fallacy of *hetu* called *varṇyasama*. This *varṇyasama* is similar to that. In both these forms of refutation, the *dṛṣṭānta*-statement which states that d has both H and S, is challenged. However, the difference between the two is that this *varṇyasama* is a false refutation because it questions the truth of any *dṛṣṭānta*-statement, assuming that it must be on par with the thesis (which is subject to doubt). Caraka's *varṇyasama* was the means of proper refutation because through it the genuine case for the dubitability of *dṛṣṭānta*-statement was made.

It may also be seen that the *jāti*-argument of the above kind is again a *prasaṅga*-argument.

§ 7.12

Avarṇyasama :

Disputant : a has S, because things having H have S. For instance d has both H and S. Similarly a has H. Thus it has S.

Opponent : According to you a is like d. But as you said, d is already proved to have S, therefore a also must be proved to have S. Thus the thesis namely a has S need not be proved by you.

Disputant : According to me a is not like d in all respects. Though d and a are similar because both are proved to have H, still they are dissimilar because d is already proved to have S, when a is not already proved to have S.

Comments : Here the opponent does not disprove the disputant's thesis but only attempts (wrongly) to show the redundance in proving the thesis.

Again, it may be seen that this *jāti*-argument also is a *prasaṅga*-argument.

§ 7.13

Vikalpasama

Disputant : a has S because the things having H have S. For instance d has both H and S. Similarly a has H. Thus it has S.

Opponent : Just as d has H, similarly e also has H, but d and e are not similar in other respects. For instance d has F but e does not have F. Thus the things having H, may not be similar in other respects. Thus a and d may not be similar in other respects. Thus d may have S and yet a may not have S.

Disputant : You have not proved the dissimilarity between d and e with respect to the predicate S. Similarly you have not proved the dissimilarity between d and a with respect to the predicate S.

§ 7.14

Sādhyasama :

Disputant : a has S because the things having H have S. For instance, d has both H and S. Similarly a has H. Thus it has S.

Opponent : According to you a is like d. Thus according to you d is like a. But 'a having S' is the thesis subject to proof. Otherwise you have to withdraw your position that a is like d.

Disputant : By accepting that a is like d, I am not accepting that a is like d in all respects.

Comments : It can be easily seen that *sādhyasama* does not differ at all from *varṇyasama*. Thus being the same as *varṇyasama*, this *jāti*-argument is also a *prasaṅga*-argument.

§ 7.15

Prāptisama :

Disputant : a has S because the things having H have S. For instance d has both H and S. Similarly a has H. Thus it has S.

Opponent : Does H prove S by reaching S? If so, then before reaching S, H does not prove S and after reaching S, H is on par with S. When H and S are on par with each other, how can we say that H is a probans and S is prabandum? It can be the other way round.

Disputant : Potter and the instruments of producing a pot can produce a pot only by reaching (i.e., coming into contact with, *prāpti*) mud, which is the raw-material. And still the distinction between them as producer and produced is retained. Thus here also the distinction between probans and probandum can be retained.

Comments : This type of *jāti*-argument is based on the wrong application of the physical model of causation to the logical model of implication. Even the disputant while answering the *jāti*-argument does not notice this wrong assumption of the opponent.

§ 7.16

Aprāptisama :

Disputant : a has S because the things having H have S. For instance d has both H and S. Similarly a has H. Thus it has S.

Opponent : Does H prove S without reaching S? If so, H cannot prove S. Because H cannot prove S unless related with S.

Disputant : When magical practices are applied for troubling the enemy, the magical forces affect the enemy without reaching the enemy. Similarly H can prove S without reaching S.

Comments : In some examples of *jāti* it may be found that the *jāti*-argument made by the opponent is not quite unsound, but the answer given by the disputant is unsatisfactory. Here also it seems that the opponent is rightly saying that *hetu* should be related with *sādhya* (though the disputant has to correct him by saying that the relation is not that of reaching : *prāpti*.) But the disputant's answer that *hetu* gives the knowledge of *sādhya* as if through some magical power, is unsatisfactory.

§ 7.17

Prasaṅgasama :

Disputant : a has S because the things having H have S. For instance d has both H and S. Similarly a has H. Thus it has S.

Opponent : If you are proving that 'a has S' on the basis that 'a has H', then you have also to prove that 'a has H' on the basis of some other reason, say, 'a has F' and so on.

Disputant : We see things in the light of some lamp (or sun etc.). But are you going to say that applying the same rule, lamp (or sun etc.) can be seen only in the light of some other lamp (or sun etc.)? Similarly here the reason that 'a has H' is explicitly true and is not in need of proof by some other reason.

Comments : Here *jāti*-argument is a wrong claim that the disputant has committed the fallacy of infinite regress. Secondly, this again is a *prasaṅga*-argument.

§ 7.18

Pratidṛṣṭāntasama :

Disputant : a has S, because the things having H have S. For instance d has both H and S. Similarly a has H. Thus it has S.

Opponent : e has H, but does not have S. Similarly, though a has H, it may not have S.

Disputant : You have not refuted my thesis because the counter-example namely e does not show that a does not have S. Thus the example namely d does not prevent the proof of the thesis.

Comments : Here the counter-example (*pratidṛṣṭānta*) that the opponent provides is the case of *vyabhicāra*. Thus the disputant has to answer, if at all he can, that it is not really the case of *vyabhicāra* (where *hetu* exists without *sādhya*). Here the disputant is unable to show it. His answer, however, that the opponent has not proved that a does not have S, is not a proper answer though it is true. Because the opponent has been successful in showing that the disputant has failed to prove that a has S.

§ 7.19

Anutpattisama :

Disputant : a is impermanent because it is a product.

Opponent : Before a is produced, it is not a product, and therefore being a non-product it is permanent.

Disputant : Before being produced, a does not exist. Therefore there is no question of a being permanent before its production. But when a is produced it is a product and hence impermanent.

Comment : This is not a general form of *jāti*-argument like the others we considered. This type of *jāti*-argument can be applied only to those cases where something is proved to be impermanent on the basis of its being a product.

§ 7.20

Śaṁśayasama

Disputant : a has S because the things having H have S. For instance d has both H and S. Similarly a has H. Thus it has S.

Opponent : Just as a and d have H, similarly they also have F. e also has F. But e does not have S, though d has S. Thus the doubt whether a has S or not continues.

Disputant : If the doubt is raised due to the similarity between a and e, namely both have F, the doubt cannot be retained because a and e are also dissimilar because a has H while e does not have H.

Comment : Akṣapāda and Vātsyāyana define *saṁśayasama* in terms of a particular example, where sound is substituted for a, the property of 'being produced through efforts' for H, impermanence for S, pot for d, universal (*sāmānya*) for e and the property of 'being perceptible through sense-organs' for F. However, the general form of *saṁśayasama* is possible to conceive of as I have done.

Secondly it may be noticed that *saṁśayasama* is very similar to *sādharmyasama* (See § 7.7) though the mode of presentation is different.

§ 7.21

Prakaraṇasama :

Disputant : a has S, because the things having H have S. For instance d has both H and S. Similarly a has H. Thus it has S.

Opponent : a does not have S because the things having F do not have S. For instance e has F but does not have S. Similarly a has F. Thus it does not have S. Now, as I can argue like this, the controversy whether a really has S or not, cannot be solved.

Disputant : While making the counter-argument you concluded that a does not have S. Thus you have tried to settle the issue. But again you said that the issue cannot be settled. Thus you are contradicting your own views.

Comments : This *jāti*-argument is very similar to *sādharmyasama* with minor differences. Vātsyāyana has also suggested that a variety of *prakaraṇasama* can be presented, which is based on the negative analogy. In that case it would be similar to *vaidharmyasama*.

§ 7.22

Ahetusama :

Disputant : a has S because things having H have S. For instance d has both H and S. Similarly a has H. Therefore it has S.

Opponent : Does H exist before S or with S or after S? If before S, then H cannot prove S in the absence of S. If with S, then H being on par with S cannot be distinguished from S. If after S, then H cannot prove S in the absence of S.

Disputant : All of us clearly see that the products are produced and the knowables are made known. You cannot question this obvious fact. S can be made known with the help of H, though H and S are not contemporaneous with each other. Secondly, your objection applies to your counter- argument also.

Comment : Here also, as in case of *prāptisama*, the opponent is wrongly applying the physical model to the logical issue.

§ 7.23

Arthāpattisama :

Disputant : a has S because the things having H have S. For instance d has both H and S. Similarly a has H. Therefore it has S.

Opponent : According to you the thesis that a has S directly follows from the premises (by force of the meaning of the premises). If so, I can provide other premises, by force of the meaning (*arthāpatti*) of which, the opposite thesis follows. a is also similar to e, because both e and a have F. But e does not have S, therefore by force of the meaning conveyed by the premises it follows that a does not have S.

Disputant : It is not sufficient for a thesis to follow by force of the meaning of its premises, that is not explicitly stated in the premises (*anukta*). (It is not the rule that whatever is not explicit in the premises must be implicitly there). But other condition that is to be fulfilled is that the respective force of the meaning (*sāmarthya*) of the premises is to be established.

Comment : Here again, the opponent's *jāti*-argument is similar to one in the case of *sādharmyasama*. The difference is in presentation because here the opponent is using the notion of *arthāpatti* as a basic notion of good inference, and is wrongly applying it to any inference whatsoever.

§ 7.24

Aviśeṣasama :

Disputant : a has S because the things having H have S. For instance d has both H and S. Similarly a has H. Thus is has S.

Opponent : You argue that a and d are non-different because both have H and then you proceed that d having S, a also should have S. In that case I can also argue that all the things in the world are non-different because all are real. Therefore as d is non-different from all other things, and d has S, therefore all other things must have S.

Disputant : The non-difference between a and d with respect to 'having S' is due to the non-difference between them with respect to 'having H', and not due to the non-difference between them with respect to 'being real'. If it is due to the non-difference with respect to 'being real' then you have to argue like this, 'All things have S because they are real'. But there you will not be able to spell out any instance which lies outside the range of the thesis. And *hetu* without instance cannot be sound.

Comments : Here again the opponent's *jāti*-argument is a *prasaṅga*-argument.

While answering the *jāti*-argument, the disputant indirectly recognises the fallacy called *anupasaṁhāri*. Though Udyotakara was first to recognise the fallacy clearly, it can thus be traced in Vātsyāyana's *NBh*.

§ 7.25

Upapattisama :

Disputant : a has S, because things having H have S. For instance d has both H and S. Similarly a has H. Thus it has S.

Opponent : If it is proper to think that a has S because a has H, then it is also proper to think that a does not have S because a has F. Thus both the reasons, 'having H' and 'having F' can be properly applied here.

Disputant : You have not refuted my thesis, as you have not denied the propriety of the reason namely 'having H', when you are saying that both H and F are proper. Now if you say that both cannot be proper because a cannot have both S and not-S, then this difficulty in again common to both H and F. Thus your objection does not disprove my thesis.

Comment : This *jāti*-argument is again clearly a version of *sādharmyasama*, with the emphasis on the term *upapatti* (= propriety).

§ 7.26

Upalabdhisama :

Disputant : a has S, because things having H have S. For instance, d has both H and S. Similarly a has H. Therefore it has S.

Opponent : Some things can be noticed (*upalabdhi*) which have S, but not H. For instance e has S, but not H. Therefore 'having H' does not imply 'having S'.

Disputant : If 'having H' is not the sufficient reason, let us think of 'having F', which is the sufficient reason, so that the thesis is established.

Comments : Here the disputant's answer is not a correct answer. The disputant should have answered that the case in which *sādhya* exists without *hetu* is not capable of refuting his argument at all. In spite of the cases where S and not-H co-exist, 'having H' can imply 'having S'. The proper case of *vyabhicāra* is of the form, '*Hetu* exists without *sādhya*' and never of the form '*Sādhya* exists without *hetu*' as the opponent supposes.

Vātsyāyana's disputant could not answer in the above manner, because as we have seen in § 2.8, the *hetu-sādhya*-relation according to Vātsyāyana was always that of 'observed equivalence'.

§ 7.27

Anupalabdhisama

Disputant : If something is not apprehended, then it is absent. a is not apprehended. Therefore a is absent.

Opponent : According to you if something is not apprehended, then it is absent. But the non-apprehension of a is not apprehended. Therefore the non-apprehension of a is absent. That is to say a is apprehended. Therefore a must be present.

Disputant : The non-apprehension of a need not be apprehended in the manner in which a needs to be apprehended for the proof of its existence. The non-apprehension is understood by itself. (*svasaṁvedya*).

Comments : This is again not a general form of *jāti*-argument, but is restricted to those cases where the disputant proves something to be absent on the basis of its non-apprehension. Secondly, this *jāti*-argument again is a *prasaṅga*- argument.

The disputant's answer is significant because it anticipates the Mīmāṁsaka doctrine that the non-apprehension of an object can generate the knowledge of the absence of that object, though the non-apprehension itself is not apprehended. Here the distinction is between the means of knowledge which generates some knowledge through its very existence (so that the knowledge of the means is not necessary for generation of the effect) and the means of knowledge which generates some knowledge only if it is itself known. (*sattayā kāraṇa* and *jñātatayā kāraṇa* respectively).

§ 7.28

Anityasama :

Disputant : a is impermanent, because a has H, and the things having H are impermanent. For instance d has H and is impermanent.

Opponent : You are arguing that a is impermanent because a is similar to d. But all the things in the world are similar to d (because all of them have the common property F). Thus it will be concluded that all the things are impermanent, which is not acceptable to you.

Disputant : My argument is not based on mere similarity but on the probans-probandum-relation noticed in the instances. If you think that only the similarity is enough, then I shall argue that your thesis is refuted, because your thesis is similar to my thesis (in the sense that both are theses). (And because my thesis is refuted by you). Secondly my hetu is not merely the positive analogy but also the negative analogy. If your *hetu* is the property which is common to all, then it cannot function as a negative analogy.

Comments : Here again the *jāti*-argument is restricted to those cases where the disputant is proving something as impermanent. This is again a case of *prasaṅga*-argument. The disputant's answer is useful for making a distinction between *anumāna* as *pramāṇa* and *jāti* (which is *apramāṇa*).

§ 7.29

Nityasama :

Disputant : a is impermanent because a has H etc. etc.

Opponent : According to you a is impermanent, which must be according to you a permanent feature of a. This means that a is permanent. (a is impermanent = a is permanently impermanent = a is permanently something = a is permanent).

Disputant : By saying that a is permanently impermanent you have not refuted my thesis that a is impermanent.

Comments : Like *anityasama*, this *jāti*-argument also a restricted to those cases where the disputant is proving something as impermanent. It can also be seen that here the opponent is quibbling on the word permanent (*nitya*). One meaning of *nitya* is 'ontologically non-changing', while the other meaning is 'undeniable'. Thus 'permanently impermanent' means undeniably impermanent, while the opponent is misinterpreting it as 'something which has the permanent (= ontologically non- changing) feature called impermanence.

§ 7.30

Kāryasama :

Disputant : Sound is impermanent b cause it is produced through efforts.

Opponent : Efforts may cause either the production of something or the uncovering of something which is covered. Thus it may be the case that sound is permanent but covered; the cover is removed through the efforts and the sound appears.

Disputant : The so-called cover of the sound is never apprehended by us. Thus the idea of uncovering of the sound through the efforts is ruled out.

Comments : This is not a general form of *jāti*, but a particular instance. In fact similar is the case with *anutpattisama, nityasama, anityasama* etc. It seems that the problem of impermanence of sound was a widely discussed problem at the time of Akṣapāda and Vātsyāyana. So they have compiled various sophistical refutations of the thesis of impermanence of sound (which was a Nyāya-doctrine) under the varieties of *jāti* and tried to answer them. In fact it is found that almost all the examples of *jāti* given by Vātsyāyana are concerned with this Nyāya-doctrine.

§ 7.31

As the nature of *anumāna* was systematically formulated in the hands of later Nyāya logicians like Jayanta and so on, the account of *jāti* also was well formulated. At the time of Akṣapāda the distinction between sound reasoning and analogical reasoning was not clear. As we saw in § 7.7, the distinction between *anumāna* and *jāti* according to them was a distinction between strong and weak analogy. As the Indian logic progressed the distinction between sound inference and analogical inference became clear and thus the most common nature of *jāti* as a type of invalid argument also became clearer in this new light.

However, though the peculiarity of most of the *jāti*-arguments was that they were based on mere positive or negative analogy and that *vyāpti* between *hetu* and *sādhya* was not asserted in them, still this cannot be the defining feature of *jātis*. As we see in the case of Vātsyāyana's *sādharmyasama* (see § 7.7) the form of the disputant's argument and the form of the opponent's counter-argument could be the same. Thus while making a *jāti*-argument the opponent may make even a valid argument with the assertion of *vyāpti*, and still such argument would be unsound. The *jāti*-arguments like *pratidṛṣṭāntasama* (see § 7.18), *anutpattisama* (see § 7.19), *ahetusama* (see § 7.22) are not genuinely based on mere similarity or difference, and a *jāti* of the type of *nityasama* is a case of quibbling (see § 7.29)

Thus it cannot be always the case that *jāti*-argument is a formal fallacy. It can even be just a material fallacy or even the so-called semilogical fallacy when it involves quibbling. Sometimes it may be found that *jāti*-argument exhibits a genuine refutation. Still Vātsyāyana always tries to show how the *jāti*-argument is unsound. Thus later on it was taken for granted that *jāti*-argument must be by definition unsound. I therefore feel that *dūṣaṇābhāsa* (pseudo-refutation) would be the most exact description of *jāti*.

§ 7.32

As we saw, many *jātis* are the improper uses of reductio ad absurdum (*prasaṅga*) method. While making a *prasaṅga*-argument the opponent shows that some unacceptable consequence follows if the disputant's argument is taken to be sound. While deriving such unacceptable consequence, the opponent generally assumes some additional premise over and above the statements made by the disputant. For instance, in the case of *aprāptisama* (see § 7.16) the opponent assumes that according to the disputant *hetu* can prove *sādhya* without reaching *sādhya*. In case of *varṇyasama* (see § 711) the opponent assumes that according to the disputant, d is like a in all respects. In such cases the disputant while answering has either to defend the additional premise and show that the additional premise is not useful for drawing unacceptable conclusion, or he has to reject the additional premise and show that his argument in itself cannot lead to unacceptable conclusion. So we find that in the case of *aprāptisama*, the disputant accepts the additional premise and tries to show that *hetu* can really prove *sādhya* without reaching *sādhya*. In case of *varṇyasama*, the disputant rejects the additional premise namely 'd is like a in all respects'.

§ 7.33

It can be seen that the earlier Buddhist logicians mostly borrowed the Nyāya account of *jāti*.[9] Diṅnāga terms *jāti* as *dūṣaṇābhāsa* and mentions fourteen types which are all borrowed from *NS*.[10] But the importance that was attached to *jāti* in Nyāya system, was not attached in Buddhist logic. The eminent Nyāya logicians always recommended the use of *jāti*-arguments in the course of hostile debates. But Dharmakīrti, an eminent Buddhist logician, asserted in *Vādanyāna* that even the debater making hostile debate ought not to use *chala* or *jāti*.[11] He, according to Dharmakīrti, should justify himself properly and refute the other debater if he can do it properly.[12] Pseudo-justifications (*sādhanābhāsa*) and pseudo-refutations (*duṣaṇābhāsa*) must not be attempted by good debaters. Dharmakīrti seems to have a greater academic spirit than any of the Nyāya logicians. Nyāya logicians preferred to behave as sophists for winning the battle of debate when Buddhist logicians like Dharmakīrti were concerned with prescribing the norms of debate.

Part III : *Nigrahasthāna*
§ 7.34

As we saw in § 1.4, Caraka has referred to many occasions of defeat and anticipated many others which were later on made explicit in *NS*. The occasions of defeat in fact exhibited the practical rules of defeat using which the judge of the debate could declare the victory or the defeat of the concerned debater. Thus occasions of defeat were basically concerned with the practical - conventional rules of debate. Some had logical import, some had merely practical import while some others had merely conventional import. In the following sections § 7.35-56 we shall consider the occasions of defeat one by one[13] and then examine their nature as compared with the nature of fallacies.

§ 7.35

Pratijñāhāni

While defining this Akṣapāda refers to the *jāti* called *pratidṛṣṭāntasama*. When the opponent refutes the disputant's argument by using this *jāti*, and if the disputant accepts the opponent's objection and admits that his argument could be unsound, then the disputant is supposed to be defeated under the occasion of defeat called *pratijñāhāni* (= giving up the thesis.)

The form of this occasion of defeat may become explicit through the following dialogue-form.

Disputant : a has S, because the things having H have S. For instance d has both H and S. Similarly a has H, thus it has S.

Opponent : But e has H, though does not have S. This counter-instance makes your thesis doubtful.

Disputant : OK, the case of counter-instance you have given may be true.

Here the disputant's answer is the occasion of his defeat.

§ 7.36

Pratijñāntara

If the disputant knows that he cannot answer the refutation made by the opponent, and shifts his thesis in order to mislead the opponent, then the disputant is supposed to be defeated under occasion of defeat called *pratijñāntara* (= different thesis).

The form of this occasion can become explicit through the following dialogue-form.

Disputant : a has S, because the things having H have S. For instance d has both H and S.

Opponent : But e has H without having S. So your thesis that a has S may be false.

Disputant : But my argument is 'a is similar to d'. Thus just as d has F, similarly a must have F.

Here the disputant, when unable to answer the opponent, shifts to a different thesis namely 'a has F'. (Here it is assumed that $S \neq F$). And thus he is supposed to be defeated.

§ 7.37

Pratijñāvirodha

If the disputant proposes some *hetu* which is contrary to the thesis itself, then the disputant is supposed to be defeated under the occasion of defeat called *pratijñāvirodha* (= contrarity with the thesis). This occasion of defeat is a fallacy which is almost identical with *viruddha* of the later Nyāya (i.e. the internal fallacy of *viruddha*). It may be remembered here that the *viruddha*-fallacy suggested by Akṣapāda and Vātsyāyana is quite different from that described by the later Nyāya logicians like Udyotakara and Jayanta.[114] The *viruddha* of Akṣapāda and Vātsyāyana is not similar to *pratijñāvirodha*.

The form of *pratijñāvirodha* may be expressed through the following dialogue-form.

Disputant : a has S, because the things having H have S. For instance d has both H and S. Similarly a has H. Thus it has S.

Opponent : In fact the things having H, do not have S at all. Thus if a has H, a must not have S. Thus 'having H' is opposed to 'having S'.

Here the opponent's refutation is accepted and the disputant is supposed to be defeated.

§ 7.38

Pratijñāsaṁnyāsa

If a disputant proposes a thesis, but withdraws it when refuted by the opponent, then he is supposed to be defeated under the occasion of defeat called *pratijñāsaṁnyāsa* (= withdrawal of the thesis). The form of this occasion may be expressed through the following dialogue-form.

Disputant : a has S because the things having H have S. For instance d has both H and S. Similarly a has H. Thus it has S.

Opponent : But ..... Therefore a may not have S.

Disputant : I am not proposing the thesis namely a has S.

§ 7.39

Hetvantara

If the disputant proposes a *hetu* for proving a thesis, but the opponent propounds that the *hetu* is not sufficient to prove the thesis, then the disputant may withdraw his first *hetu* and propose another *hetu* that the disputant feels to be sufficient. When such another *hetu* is proposed by the disputant he is supposed to be defeated due to *hetvantara* (= different *hetu*).

The form of this occasion may be clear from the following dialogue-form.

Disputant : a has S for the things having H have S. For instance d has both H and S. Similarly a has H. Therefore it has S.

Opponent : ..... Thus H is not sufficient to prove S.

Disputant : OK, then I shall argue like this. a has S because the things having F have S. For instance e has both F and S. Similarly a has F. Therefore it has S.

§ 7.40

Arthāntara

If the disputant talks about irrelevant matters in the debate, then the disputant is supposed to be defeated under the occasion of defeat called *arthāntara* (= different, i.e., irrelevant contention).

For instance, if a disputant presents his thesis and argues for it, and while justifying the argument explains how the words used in the argument are grammatically correct (when the opponent has no objection regarding this), then the grammatical matters being irrelevant, the disputant is supposed to be defeated. 'God exists because 2 + 2 = 4' is also an example of this type.

§ 7.41

Nirarathaka

The disputant who commits *arthāntara*, speaks something meaningful though irrelevant. But the disputant who commits *nirarthaka*, simply utters meaningless sounds in the course of justification of his thesis. *Nirarthaka* means meaningless. For instance, somebody may seem to argue like this. 'Sound is eternal. ABCD, because EFGH, like XYZ'.

§ 7.42

Avijñātārtha

If some sentence of the disputant, although uttered thrice, is not understood either by the opponent or by anybody else in the council, then the disputant is supposed to be defeated under the occasion of defeat called *avijñātārtha* (something, the meaning of which is not understood).

Generally the disputant may play a trick of uttering an obscure statement for hiding his incapability to justify himself properly. The disputant for this sake may twist on some words, or may make unconventional use of certain words, or utter the statement so fast that nobody should grasp it,[15] and so on.

§ 7.43

Apārthaka

The debater who commits *apārthaka*, does not utter just meaningless sounds like the one who commits *nirarthaka*. Though *apārthaka* is much similar to *arthāntara*, because in both the cases the disputant talks irrelevant things, still there is some difference between the two. In the case of *arthāntara*, the irrelevant statements of the disputant make sense. They are internally related and have some (may be indirect) connection with the disputant's argument. They cause the occasion of defeat because they are irrelevant to the issue in hand. But the statements that the disputant makes while committing *apārthaka*, are not only irrelevant to the issue in hand, but also lack any mutual connections or any connections with the argument made. These statements are absurd utterances, though not meaningless in themselves. They are meaningless in connection with eath other. For instance somebody may 'argue' like this. 'Sound is impermanent. And there are ten pomegranates, six cakes, a vessel, the skin of a goat, a lump of sesamum, and the lower part of the thigh. You have to make the girl drink it. Her father is not cold'.

§ 7.44

Aprāptakāla

According to the Nyāya logicians the number of the necessary and sufficient members (*avayava*) of the inferencial statement was five. According to them not only the number of the members but also the order in communicating the members was important. If, while stating the inferencial statement, the disputant presents the members in a different order (as compared with the order standardised by Nyāya) then the disputant was supposed to be defeated under the occasion of defeat called *aprāptakāla* (= mistimed). The standard order of the five members was as follows.

1. *Pratijñā* : The hill is fiery.

2. *Hetu* : Because the hill is smoky.

3. *Dṛṣṭānta* : Like the kitchen. (Or : Whatever is smoky is fiery like the kitchen)

4. *Upanaya* : Analogously the hill is smoky.

5. *Nigamana* : Therefore it is fiery.

The change in the above order was regarded as an occasion of defeat.

§ 7.45

Nyūna

Nyāya logicians emphasised that the number of the members of the inferencial statement should be exactly five. Thus if the disputant while arguing presents less number of members as the members of his inferencial statement, then it is regarded as the occasion of his defeat called *nyūna* (= insufficient).

§ 7.46

Adhika

If while stating the inferencial statement a disputant presents more than one *hetus* or more than one instances (or both), where only one *hetu* and one instance are sufficient for proving the thesis, then the disputant is supposed to be defeated due to *adhika* (= excessive). Here the statements that the disputant while committing *adhika* is adding to the necessary and sufficient number of the members of the argument, are not false, or logically inadequate, but they are superfluous.

§ 7.47

Punarukta

Repeating one and the same sentence or repeating one and the same proposition by using synonymous expressions was supposed to be an occasion of defeat called *punarukta* (= iteration). If some statement of the disputant is not understood by the opponent as well as by the members of the council, then according to the rules of debate the disputant has to repeat the statement maximally twice. (In that case the disputant is not supposed to commit *punarukta*, though he iterates.) And if the opponent and the members of the council still fail to understand the statement, then the disputant is supposed to have committed *avijñātārtha*. But if the disputant repeats his statement unnecessarily (i.e. in the situations other than that are described above), the disputant is supposed to have committed *punarukta*.

§ 7.48

Ananubhāṣaṇa

We saw in case of *avijñātārtha* (see § 7.42) that if some statement of the disputant is not understood by any member of the council, even though it is uttered thrice, then the disputant is supposed to be defeated. Sometimes, however, it may happen that some members of the council are able to understand the disputant's statement, while the opponent cannot, or though he does, he cannot restate the disputant's statement. In that case the opponent is supposed to be defeated due to *ananubhāṣaṇa* (= no restatement).

§ 7.49

Ajñāna

In the case of *ananubhāṣaṇa* the opponent may have understood the disputant's proposition, but he is unable to respond to it. But in the case of *ajñāna* (=failure to understand) the opponent does not understand the disputant's proposition at all. Here also the disputant has to utter the statement thrice and some members of the council must understand the statement, though the opponent does not. If on the contrary, all the members of the council also fail to understand, then it is not the opponent who is defeated due to *ajñāna*, but it is the disputant himself who is defeated due to *avijñātārtha*.

§ 7.50

Apratibhā

As in the case of *ananubhāṣaṇa* (see § 7.48), also in the case of *apratibhā* (= lack of imagination), the opponent is unable to answer the disputant's argument. But the difference between the two is that in case of *ananubhāṣaṇa*, the opponent being unable to answer or even to restate the disputant's position, keeps mum, while in case of *apratibhā*, the opponent may express his inability to imagine the correct answer.

§ 7.51

Vikṣepa

The opponent who is unable to understand or answer the disputant's argument, may not decide either to keep mum or to express the inabilities, but he may try to distract the disputant and say that he has just now got some other work to do, that is why he cannot continue with the discussion. The debater who is asking for such excuses is supposed to be defeated due to *vikṣepa* (= distraction).

§ 7.52

Matānujñā

When the opponent refutes the disputant's position, the disputant who is ubale to answer the opponent's refutation may try to refute the position of the opponent. Here, as the disputant does not give any direct answer to the objections raised by the opponent, the disputant is taken to have accepted the opponent's criticism and thus is supposed to be defeated due to *matānujñā* (= admitting the opinion(s).)

§ 7. 53

Paryanuyojyopekṣaṇa

It may happen in a course of debate that there comes an occasion of the disputant's defeat, but the opponent fails to recognise that. Here the disputant should be taken as defeated by the opponent, but the opponent continues with his own argument. In that case as the opponent fails to notice the fault of the disputant, he, rather than the disputant is to be declared as defeated by the council. This occasion of defeat is called *paryanuyojyopekṣaṇa* (= neglect of the occasion of (other's) defeat).

§ 7.54

Niranuyojyānuyoga

Sometimes, as a contrary case of *paryanuyojyopekṣaṇa*, it may happen that though there is not any occasion of defeat on part of the disputant, the opponent may make a (false) claim of some occasion of the disputant's defeat. In such cases the council obviously has to declare the occasion of the opponent's defeat. This occasion of defeat was called *niranuyojyānuyoga* (= claiming an occasion of defeat when there is none.)

§ 7.55

Apasiddhānta

In the course of debate, a debater may make a proposition or state a thesis which is contrary to the position he has already accepted. In that case the debater is supposed to be defeated due to *apasiddhānta* (= thesis incompatible with one's own accepted position). This occasion of defeat is very similar to the *viruddha* fallacy as stated by Akṣapāda and Vātsyāyana. The only difference is that in the case of *viruddha* fallacy it is the *hetu* presented by the debater which is incompatible with the accepted position of the debater, while in case of *apasiddhānta*, it is the thesis or the proposition which is so incompatible.

§ 7.56

Hetvābhāsa

All the fallacies of *hetu* discussed by Akṣapāda and Vātsyāyana were regarded by them as the occasions of defeat.

A question can be raised here as to why Akṣapāda has included *hetvābhāsa* in the types of the occasions of defeat, when he has mentioned it separately also in his list of the important concepts[16] in the methodology of debate. The answer is that though *hetvābhāsas* are significant as the occasions of defeat, they are not significant only in the context of hostile debate as 'the occasions of defeat' are. The notions of victory or defeat are applicable only to hostile debate, but the notion of *hetvābhāsa* is significant even in a friendly debate. In friendly debates *hetvābhāsas* may be committed, though they would not be regarded as the occasions of defeat. The basic distinction that Akṣapāda could have in his mind here is that between logical/philosophical faults and the practical/conventional faults. However, we shall see that many occasions of defeat other than *hetvābhāsa* are on par with *hetvābhāsa* in this respect.

§ 7.57

Dharmakīrti in his treatise called *Vādanyāya* refines and systematises Akṣapāda's list of the occasions of defeat. According to his classification there are four basic types of the occasions of defeat.

1) Failure to give necessary and sufficient justification of one's own thesis.

2) Making superfluous or faulty utterances/statements/ arguments in the course of debate.

3) Failure to point out the faults (which include fallacies and other occasions of defeat) committed by one's opponent.

4) Claim that one's opponent has made a certain fault when the opponent has actually made none.

The first two types in the above classification are suggested by the two alternative derivations of the complex term '*asādhanāṅgavacana*' while the other two types are suggested by the two alternative derivations of the complex term '*adoṣodbhāvana*; used by Dharmakīrti.[17] Many of the occasions of defeat discussed by Akṣapāda can be brought under these four types. Thus, *pratijñāhāni*, *pratijñāntara*, *pratijñāsannyāsa*, *hetvantara*, *nyūna* (in Dharmakirti's sense), *ananubhāṣaṇa*, *ajñāna*, *apratibhā*, *vikṣepa* and *matānujñā* appear as the subvarieties of the Dharmakīrti's first type. One's committing any of these amounts to one's implicit or explicit failure to give sufficient justification of one's own position. *Pratijñāvirodha*, *apasiddhānta*, *hetvābhāsa*, *arthāntara*, *nirarthaka*, *avijñātārtha*, *apārthaka*, *punarukta* and *adhika* appear as the subvarieties of the Dharmakīrti's second type. First three of these are the explicit logical/philosophical faults, while the remaining amount to some superfluous utterance/statement/argument.

However, *pratijñāntara*, *hetvantara*, *vikṣepa* and *matānujñā*, which we have included in first type, also amount to superfluous statement/argument and thus can be regarded as a subvariety of the second type as well. Similarly making superfluous utterances may sometimes be an implicit indication of the failure to give sufficient justification. Thus there is no water-tight compartment between the first and the second type.

Paryanuyojyopekṣaṇa of Akṣapāda seems to be identical with Dharmakīrti's third type. And *niranuyojyānuyoga* of Akṣapāda seems to be identical with Dharmakīrti's fourth type.

Accroding to Dharmakīrti, *aprāptakāla* is by no means an occasion of defeat. Thus it cannot be included in any of the Dharmakīrti's types of the occasions of defeat.

§ 7.58

Dharmakīrti accepts *nuūna* as an occasion of defeat only in a restricted sense. If in an inferencial statement the so-called member *(avayava) pratijñā* or *upanaya* or *nigamana* is lacking, then unlike Nyāya logicians, Dharmakīrti would not regard this as a case of an occasion of defeat called *nyūna*. For according to Dharmakīrti *pratijñā*, *upanaya* and *nigamana* are not at all the necessary members of an inferencial statement. For him only two statements are necessary and sufficient : the statement that states that *hetu* exists in *pakṣa*-case and the statement that states the *vyāpti* between *hetu* and *sādhya*. If any of these two statements is lacking, then making such insufficient inferencial statement amounts to *nyūna* according to Dharmakīrti.

As against the Nyāya logicians, Dharmakīrti suggests that stating *pratijñā* or *upanaya* or *nigamana* in the course of the proof of a thesis, itself is an occasion of defeat because the debater who is stating any of these is making a superfluous statement.[18] Though this position of Dharmakīrti looks sound in the context of Dharmakīrti's general frame of debate, as far as *upanaya* is concerned, it seems, however, that either *pratijñā* or *nigamana* is to be regarded as a necessary part of the inferencial statement and not a superfluous one.

Here, it may be argued that the debater who wants to prove certain thesis has to clarify what he is trying to prove. Thus either *pratijñā* (= claim of the thesis) or *nigamana* (= conclusion) is a necessary part of the total argument. Though it is true, as Dharmakīrti holds, that logically the conclusion of a sound inference need not be stated as it logically follows from the premises, still a genuine practical problem in any debate is that we can conceive of a number of conclusions which logically follow from any given set of premises. So the conclusion which the debater is intending to draw from the premises he is stating, may not become clear to the other debater unless it is verbally stated by the arguing debater. We have to distinguish here between the genuinely necessary parts of an argument made in a debate, and the necessary parts of 'proof'. We have further to distinguish between an argument (= '*parārthānumāna*' *of Diṅnāga*) *and a proof* (= '*sādhana*' of Dharmakīrti). Though Diṅnāga used the terms '*parārthānumāna*' and '*sādhana*' as synonyms, he included the statement of *pakṣa* as a part of *parārthānumāna* (or *sādhana*). Thus Diṅnāga's notion of *parārthānumāna* was that of a full-fledged argument with *pakṣa, hetu* and *dṛṣṭānta*. According to Dharmakīrti *pakṣa* was not a part of *sādhana*. Thus Dharmakīrti's notion of *sādhana* was that of a proof.

For instance if we have an argument of the form (A) :

Step 1 : Wherever there is smoke, there is fire.

Step 2 : There is smoke on the mountain.

Therefore

Step 3 : There is fire on the mountain.

We may say that the first two steps provide the proof of the thesis stated in step 3. In this sense step 3 is not the part of the proof (= *sādhana*) of the thesis. But step 3 is the part of the full-fledged argument (A). The argument (A) can be presented in reverse manner also where we have, (B) :

Step 1 : There is fire on the mountain

Because

Step 2 : Wherever there is smoke, there is fire

And Step 3 : There is smoke on the mountain

Although step 3 of the argument (A) and Step 1 of the argument (B) are the same in content, in (B) its position is that of *pratijñā* while in (A) its position is that of *nigamana*. Thus either *pratijñā* or *nigamana* is to be there in any full-fledged argument. The debater has not only to present the proof of some implicit thesis, but he has also to make his intended thesis explicit.

§ 7.59

An important question regarding the methodology of debate is, what should be the criteria for determining the success or failure of a debater. A simple view that can be held here is that a debater wins if his position is faultless or less faulty than that of the other debater. According to this view the soundness or the comparative strength of a position becomes the criterion of the success of the debater. However, practically it is the manner of presentation that plays the major role. When the debater A while presenting and justifying his thesis commits fallacies, his opponent, the debater B may be deceived by A's manner of presentation and may be unable to point out the fallacies committed by him. In such a case, the debater B is defeated and A, whose argument is fallacious wins. (See the case of *paryanuyojyopekṣaṇa*, § 7.53). Here, where the presentation of A's position has deceived B, A may have made his deceptive arguments purposefully, when he is not able to justify his own position in a legitimate way.

We see that Nyāya authors explained and recommended the tools like *chala* and *jāti* for the sake of deceiving or misleading the other debaters and protecting one's own position. Though Dharmakīrti was opposed to the use of *chala* and *jāti* in course of debate, he did not reconsider the above-mentioned rule that if the fallacy committed by A is not noticed by B, then A wins and B is defeated. (See Dharmakīrti's third occasion of defeat mentioned in § 7.57) Thus the frame of debate accepted by Dharmakīrti too was loose enough to allow the 'successful' use of *chala* and *jāti*.

§ 7.60

In ancient Indian treatises on logic, hostile debate was discussed like a serious game and like any other games it had rules. We see that there are some rules of a game which are flexible. By changing those rules we can have different subvarieties of the same game. These rules in the context of a particular subvariety of the game can be fixed according to convenience and conventions of those who are playing the game. But we also see that some rules of the game are basic to that game. The change in these basic rules amounts to the change in the game itself. Accordingly we can talk of basic rules of debate and flexible rules of debate. Breaches of the basic rules of debate may be called *categorically proper* occasions of defeat, while breaches of the flexible rules of debate may be called *conditionally proper* occasions of defeat. In the case of some rules of debate it can be doubted whether they are proper (basic or flexible) rules at all. Such rules may be rooted in the false prejudices of the theoreticians of debate, regarding the logic and philosophy of debate. The breaches of such improper rules shall be called improper occasions of defeat.

§ 7.61

The basic rules of the game of debate regarding the success or the failure of the debater seem to be as follows -

1. Any debater should be able to prove and justify the thesis he has presented in the course of debate. While doing this he should not commit any fallacies. Otherwise he may be taken to be defeated.

2. Any debater should not make any meaningless or irrelevant utterances. Otherwise he may be taken to be defeated.

The flexible rules seem to be like the following. (I shall enumerate only some) .

1. Any debater should not repeat any of his sentences unless asked by the other debater or the judge or the council.

2. Any debater should be able to point out the fallacy committed by his opponent, immediately after it is committed.

3. Any debater should not propose more than one *hetus* for proving one and the same thesis.

The improper rules of debate can be like the following (I shall enumerate only a few of them)

1. Every inferencial statement posed by any debater shall contain five members.

2. The order of presentation of the members of the inferencial statement should be : *pratijñā, hetu, dṛṣṭānta, upanaya, nigamana.*

§ 7.62

The Nyāya-occasions of defeat, namely *pratijñāvirodhā, nyūna* (with restrictions, see § 7.58), *niranuyojyānuyoga, apasiddhānta* and *hetvābhāsa* are fallacies. Thus committing any of these fallacies amounts to a breach of the basic rule (1). These can therefore be called categorically proper occasions of defeat. But even in that case we can conceive of a game of debate where the debater who has committed a fallacy is given some chance for giving an alternate non-fallacious proof of his position.

If a debater when objected by his opponent keeps mum, or expresses his failure to understand the objection (though it is understood by all others), expresses the lack of imagination to think of the good answer, or withdraws his own thesis, then he is in fact unable to justify himself and thus breaks the basic rule (1). Accordingly *ananubhāṣaṇa, ajñāna, vikṣepa, apratibhā, pratijñāhāni* and *pratijñāsaṁnyāsa* can be called categorically proper occasions of defeat. Similarly the debater committing *pratijñāntara* or *hetvantara* has in fact withdrawn his first argument and thus has broken the basic rule (1). But if in the game of debate the second chance for providing a sound proof of the same thesis is offered, then *hetvantara* may not always be regarded as categorically proper occasion of defeat, because the second *hetu* proposed by the debater may be sound. But the same thing does not apply to *pratijñāntara* where the debater shifts his thesis itself.

Nirarthaka is the case of meaningless utterances while *arthāntara* is the case of irrelevant statements. *Apārthaka* is the case of both. Thus these being the breaches of the basic rule (2), can be called categorically proper occasions of defeat. In fact the breaches of the basic rule (2) are suggestive of the breaking of the basic rule (1) because generally the debater who uses *arthāntara* or *nirarthaka* or *apārthaka*, tries to hide his inability to justify his own thesis properly.

§ 7.63

In some cases of the Nyāya-occasions of defeat, it becomes disputable whether they really break some basic rule or not. For instance, the debater committing *hetvantara* provides another *hetu* for proving the same thesis. If his second *hetu* sufficiently proves the thesis, it would be an injustice to charge him of the occasion of defeat called *hetvantara*. It is true that the debater committing *hetvantara*, has in a way accepted that the first *hetu* provided by him is insufficient. But we can conceive of a good game of debate, where any disputant is given a specific number of chances for providing an appropriate *hetu*. In Nyāya and even in Buddhist frame (and for that matter in any ancient Indian frame) of debate such chances were not offered to any debater. However the presupposed rule here, namely, no chances should be given to any debater for providing a better argument, is a flexible rule of debate and not a basic one.

§ 7.64

The Nyāya-occasion of defeat namely *avijñātārtha* is also disputable in a similar way. On several occasions it may happen that the debater who uses obscure expressions and thus is not understood by the other debater or by the council, may do so for hiding his inability to justify his thesis properly. But this may not be always the case. Sometimes the thought expressed by the debater may be such that it cannot be easily expressed by him without using such obscure expressions or technical jargons. In that case it is difficult to judge whether the use of obscure expressions really breaks the basic rule (1). We can conceive of a good game of debate where the disputant who uses such expressions is given a specific number of chances for expressing the same thought in simpler words. But in ancient frame of debate, it seems, no such chances were offered. We find that three chances (Why three ? And not two or four ? Again this is a matter of convenience and not a basic rule.) were offered to repeat (but not to restate) the obscure statement so that it may be understood by others.

§ 7.65

The Nyāya-occasion of defeat called *aprāptakāla* seems to be an improper occasion of defeat because keeping a particular order in presenting the members of an inferencial statements is not at all necessary. (The only rule that seems legitimate is that the conclusion should be placed immediately before or after the premises and not in the middle of the premises. But no rule can be made regarding the order of the premises). But this order (even of the premises) was an accepted convention at the time of Akṣapāda and breaking this convention was supposed to be an important mistake. In fact it was thought by the Nyāya logicians that the particular order of the members of the inferencial statement had a philosophical significance. It was thought that coming to know the conclusion of an argument was an outcome of a process and for giving the result namely 'the knowledge of the conclusion,' the process must be a well-ordered process. This was the process of knowing an argument. Corresponding to the order in the proper process of knowing an argument, it was thought, there should be the order in the process of making that argument. So if the conclusion of one's argument is to be well understood by others, one has to follow a particular order of the members of argument while making the argument. Further it was thought by Nyāya logicians that there was some isomorphism between *svārthānumāna* and *parārthānumāna*. (The advocates of *parāmarśa* seem to be responsible for this). In the case of *svārthānumāna*, the inferring agent sees as first step, *hetu* on *pakṣa*.

Corresponding to it there was the member called *hetu* in *parārthānumāna*. (It may be noted that *pratijña* of *parārthānumāna* has no corresponding step in *svārthānumāna*). Similarly the next three steps in *svārthānumā* namely 1) recollection of *vyāpti* 2) *parāmarśa* and 3) *anumiti* corresponded to the next three members of *parārthānumāna* namely 1. *dṛṣṭānta* 2. *upanaya* and 3. *nigamana* respectively. Thus the order in the members of *parārthānumāna* was justified by Nyāya logicians as being isomorphic with the order in the steps of *svārthānumāna*. However the isomorphism between *svārthānumāna* and *parārthānumāna* itself is questionable. Because *svārthānumāna* need not be always a complex cognitive process. Immediately after one sees *hetu* on *pakṣa*, one may infer *sādhya*. This inference is like an intuition. But for justifying the truth of that intuitive knowledge, one may provide the premises like *vyāpti*-statement etc.. Thus the way of knowing and the way of justifying may not be isomorphic at all. Buddhists seem to have seen this point.[19]

Secondly the specific order in the premises of an argument is neither logically (because the relation between the premises is that of conjunction and conjunction is commutative) nor practically (because we do understand arguments in spite of the change in the order of pemises) necessary. This is how, Dharmakīrti allows the changes in the order of the premises.[20]

§ 7.66

There is a problem regarding *nyūna* also. We can see that out of the five members of inferencial statement as accepted by Nyāya, only three members are the logically necessary parts of a sound argument. They are *hetu* (which is a *pakṣadharmatā*-statement) *dṛṣṭānta* (which contains universal *vyāpti*-statement) and either of *pratijñā* and *nigamana*. As we saw in § 7.58 and and elsewhere, Dharmakīrti accepted only first two of the above-stated three, though he could have accepted the third also. *Upanaya* in fact served only as the conjunction of *hetu* and *dṛṣṭānta*, and making a separate conjunctive statement called *upanaya* for serving this purpose was not at all necessary because this purpose could be served by stating *hetu* and *dṛṣṭānta* in succession. Secondly if any of the two, namely *pratijñā* and *nigamana*, is stated, the remaining member is redundant, as far as it does not serve any additional logical purpose. (Again see § 7.58)

Thus we may conclude that *nyūna* may be regarded as a proper occasion of defeat only if the debater does not and cannot present logically necessary parts of the argument he is trying to make. The debater may be even allowed to present enthymemes, if the other debater understands and accepts the suppressed premise (s) and if the arguing debater is able to state and justify the suppressed premise(s) in a sound manner. In this sense, *nyūna* as a fallacy should be distinguished even from enthymeme. Or if *nyūna* is identified with enthymeme, then *nyūna* (even in Dharmakīrti's sense) should not be always regarded as a fallacy.

§ 7.67

We have a greater problem with respect to the Nyāya-occasion of defeat called *adhika*. Because giving an additional *hetu* and additional instance in the course of proof is neither necessarily making the argument unsound; nor is it irrelevant, or meaningless. Thus the rule of debate that a debater should provide in an argument one and only one *hetu* and one and only one instance is too strict, and need not be regarded as a basic rule. However it may be a flexible rule of some variety of debate.

§ 7.68

In the case of the Nyāya-occasion of defeat called *punarukta* we have the problem similar to one that arises with respect to *avijñātārtha*. The debater may sometimes repeat himself for hiding his inability to give proper proof. But sometimes the debater, without having such intention, may habitually repeat his statement. Sometimes the debater may do so for making himself easily understandable. Thus it is difficult to judge whether the repetition really amounts to breaking of the basic rule (1). The basic question seems to be whether in spite of repeating himself, the debater justifies himself soundly or not. In this way Dharmakīrti's second type of the occasion of defeat is also disputable because sometimes superfluity may not be a serious flaw.

§ 7.69

The Nyāya-occasion of defeat called *matānujñā* also is problematic. When the disputant, instead of answering the refutation made by the opponent, criticises opponent's own position, he contends that the faults similar to those in his position can be shown in the opponent's position as well. In such a case the disputant's position and the opponent's position remain on par and the opponent's refutation of the disputant's position can not be held conclusive. A verse that is often cited in Nyāya-literature states, "When the two (opposite) positions have similar drawbacks and, may be, also the similar ways of removing the drawbacks, then while discussing such an issue any party should not be taken as refuted."[21]

Sometimes the disputant may choose to refute the opponent's counter-thesis for indirectly defending his own thesis. This indirect method of self-defence was many times followed by many eminent philosophers. Though this method seems to be rejected by Akṣapāda from the procedure of a good hostile debate when he introduced *matānujñā*, it cannot be always the case that the so- called situation of *matānujñā* is always the one in which a basic rule of debate is violated.

§ 7.70

Lastly the Nyāya-occasion of defeat called *paryanuyojyopekṣaṇa* is objectionable. (See § 7.59). Sometimes the debater A without pointing out the faults in B's objections may show how his thesis remains in tact even though such objections are raised. And thus A may proceed with his constructive arguments. In such a case it would be proper to be seen by the judge and the council whether the arguments made and the answers given by A are sound or not. Thus the situation of *paryanuyojyopekṣaṇa* cannot always amount to breaking the basic rule (1).

Sometimes, however, it may happen that A can neither refute B's objections, nor answer them. In such a case it would mean that A cannot justify himself properly and thus it would be a proper occasion of his defeat. Though such an occasion can be classified under *paryanuyojyopekṣaṇa*, it can be regarded as a case of *apratibhā* or *ananubhāṣaṇa* as well. (See §§ 7.48,50)

Part IV : Prasaṅga
§ 7.71

In the ancient Indian treatises on logic we do not find the discussion on *prasaṅga*-fallacies as a part of the discussions on fallacies in general. Though some of the *prasaṅga*-fallacies are indirectly considered in the discussions on *jāti*,[22] still many important *prasaṅga*-fallacies remain undiscussed. *Ātmāśraya, itaretarāśraya, cakraka* and *anavasthā* are such important *prasaṅga*-fallacies. They are used by various Indian philosophers for both proving one's own position as also for refuting the opponent's position. But we do not find them explained or discussed in any ancient orthodox treatise on logic.

Udayana is perhaps the first Indian logician to include these fallacies under *tarka*. But this *tarka* as described by Udayana differs from *tarka* described by the ancient Nyāya logicians. In ancient Nyāya starting from Akṣapāda to Bhāsarvajña and Vācaspati, *tarka* was described as a speculation regarding the truth of the proposed thesis. This *tarka* can be regarded as a cognition which is more certain than a doubt *(saṁśaya)* but less certain than certainty *(niścaya)*. Doubt may be of the form 'p or q' (where p and q stand for propositions) while certainty may be of the form p (or even of the form 'Necessarily p'). *Tarka* in that case which is a speculation regarding the truth of p, is the cognition of the form 'Most probably p.' This *tarka* helps us in attaining certainty, but is not itself certain. Now how is this speculative cognition of *tarka* obtained? Generally it was claimed that this cognition was obtained by discussing as to which thesis, whether p or q seems to have a strong support of causal laws.[23] Thus the ancient method of using *tarka* very much differed from the *prasaṅga*-method. Of course, though not under *tarka*, *prasaṅga*- method was in use long before Udayana, and many *prasaṅga*-fallacies find mention in the treatises of ancient Indian logic.

§ 7.72

The method of *prasaṅga* may be used as a method of proof or of disproof. *Prasaṅga*-method as a method of proof was in fact the method of indirect proof. It is seen in Indian logic that the method of indirect proof was generally given a lower status than the method of direct proof. The method of indirect proof was, so to say, not regarded as a genuine method of proof at all,[24] but only a method by which the direct proof was strengthened. That is also the reason why *tarka* was never regarded as a *pramāṇa* (=means of authentic cognition) by the Nyāya logicians. However, the method of *prasaṅga* formally tallies with the method of reductio ad absurdum. Following are the patterns of proof and disproof by *prasaṅga*-method.

Let us understand by t, the thesis to be proved by *prasaṅga*-method and by t́, the thesis to be disproved by *prasaṅga*-method. Let us understand by u an unacceptable proposition.

1) Proof by *prasaṅga*-method.

The thesis : t

Proof : Let us suppose that t is false

Then u

But u is unacceptable.

Therefore t is true.

2) Disproof by *prasaṅga*-method.

The thesis : t́

Disproof : Let us suppose that t́ is true

Then u

But u is unacceptable.

Therefore ť is false.

In both the above patterns u indicates an unacceptable proposition. There are many kinds of unacceptable propositions. For instance u may be a contradictory statement, or may refer to a circular position, or to infinite regress and so on. Sometimes u may be contrary to some other statement accepted by the same debater who has proposed the thesis.

While proving the thesis t by *prasaṅga*-method one does not show any *prasaṅga*-fallacy in the argument, but one argues that if any counter-argument to disprove t is made, then it would lead to u and thus would involve a *prasaṅga*-fallacy. While disproving the thesis ť by *prasaṅga*-method, however, one shows that any argument made for proving ť involves a *prasaṅga*-fallacy.

The term *prasaṅga* is derived from pra + sañj and means 'being forced to accept something.' Generally this 'something'is a calamity or an unacceptable situation. Thus *ātmāśrayaprasaṅga* would mean 'being forced to accept something's dependence on itself.' Here 'something's dependence on itself' is the calamity, an unacceptable situation. Similarly the terms *itaretarāśrayaprasaṅga*, *cakrakaprasaṅga* (popularly known as *cakrakāpatti)*, *anavasthāprasaṅga* etc. can be explained.

In the following sections § 7.73 - 79 we discuss the important *prasaṅga*-fallacies one by one.

§ 7.73

Atmāśraya (Self-dependence)

In the case of *ātmāśraya*, u (= unacceptable proposition) is of any of the following forms.

1. X depends on X for its being known by anybody.

2. X is the substratum of X.

3. X is the object of X.

4. X is produced out of X.

We can conceive of many such forms where X, though the same as X, is predicated of X somehow or the other. The error underlying such a proposition is that while establishing any relation (other than identity) between X and X we forget that X cannot be different from itself. In this way *ātmāśraya* is a logical error. In the following dialogue we may get an example of this fallacy.

Cārvāka : Consciousness appears through the matter itself and is not separate from the matter.

Śaṅkara What is the nature of that consciousness which according to you appears through the matter itself?

Cārvāka It is the experiencing of the matter, which is called as consciousness.

Śaṁkara : (You are committing the fallacy of *āmāśrayaprasaṅga*). According to you then matter experiences itself. But the same matter cannot be the object of itself. Nothing can act on itself. For instance the fire, though burning, cannot burn itself. Any stage actor, however trained he may be, cannot climb on his own shoulders.[25]

Here what Śaṅkara is doing is that he is pointing out that any argument attempted to prove the thesis namely 'Consciousness arises when matter experiences itself' would be fallacious due to *ātmāśrayaprasaṅga*. Though in the actual discussion in *BSSB* Śaṅkara has not used any synonym of *ātmāśraya*, his analogies of the fire and the stage-actor are pointers of the same.

§ 7.74

Itaretarāśraya (Mutual dependence)

In the case of *itaretarāśraya* (also called as *anyonyāśraya*) our u is of any of the following forms.

1) X depends on Y for its own existence and Y depends on X for its own existence. (Such a situation may not always be unacceptable, but the fallacy would occur only if it is 'unacceptable.')

2) The knowledge of X depends on the prior knowledge of Y and vice versa.

Thus we can conceive of two kinds of the mutual dependence, existential and epistemic. Even the epistemic can be reduced to existential when we accept knowledge to be an entity and consider the causality between one knowledge and another knowledge. In the following dialogue we may get an example of the fallacy.

Disputant	The criterion for judging that a means of cognition is *pramāṇa* is that with the help of it we should get true cognition of the object *(=prameya)*
Opponent (Nāgārjun)	How to be sure that the cognition of the object is true ?
Disputent	It is judged by the authenticity of the means of cognition.
Opponent (Nāgārguna)	Then you are led to *itaretarāśraya* because according to you the knowledge of *pramāṇa* as *pramāṇa* depends on the knowledge of *prameya* (= the object of true cognition) as *prameya* and vice versa. You are as if advocating the thesis that the father gives birth to the child and the child gives birth to its father.[26]

What Nāgārjuna is trying to point out is that if *pramāṇa* and *prameya* are established on the basis of each other, then they depend on each other for being known, and thus we are committing *itaretarāśraya*. Though Nāgārjuna, in *Vigrahavyāvartanī* does not use any word which is a synonym of *itaretarāśraya*, his intention is clear. It can also be seen that Nāgārjuna considers the two types of *itaretarāśraya* (namely existential and epistemic) on par when he compares the case of epistemic *itaretarāśraya* with the causal mutual dependence between a father and his child.

§ 7.75

Cakraka (circular dependence)

I have translated *cakraka* as circular dependence to distinguish it from circularity, (or arguing in a circle), the popular fallacy (see § 8.40). The difference between *itaretarāśraya* and *cakraka* is not a categorical difference but the difference in degree. In the case of *cakraka* there is a chain of dependence (again existential or epistemic) between more than two entities while in the case of the first the depending and depended entities are only two. For instance while committing *cakraka*, it may be argued that A depends on B, B on C, C on D and D on A. The following dialogue will provide an example of *cakraka*.

Disputant	In the beginning of *Carakasaṁhitā*, the author mentions the purpose of writing the book, so that the reader could be induced to read whole of the book.
Opponent	How can the reader know that the prupose of the book is served by the book just by reading the purpose of the book stated by the author ?
Disputant	He could know it because he could have known that the author is an authentic person.
Opponent	How would the reader know that the author of the book is an authentic person ?
Disputant	He could easily conclude this when he reads the book.
Opponent	But how could he read the book unless he is induced to read it? How could he be so induced unless he knows that the purpose of the book is served by it? How could he know that the purpose is served unless he knows that the author is authentic ? And how can he be sure about the authenticity of the author unless he reads the whole book? Thus your argument is led to circular dependence.[27]

§ 7.76

Anavasthā (Infinite regress)

In the the case of this *prasaṅga*-fallacy, our u is of any of the following forms.

1. There are infinite number of things/objects having such and such character, in the world.

2. Infinite number of events of such and such character should happen before/after such and such event happens.

We shall see in § 7.77 that the statement of the above form may not be false every time. Thus we have to check in any concrete case of the so-called *anavasthā* whether the so-called 'u' there is really an unacceptable proposition. In this section we shall consider two genuine examples of *anavasthā* as a *prasaṅga*-fallacy. The following two dialogues will provide the examples.

(I) Disputant The atoms and the molecule which is made of the atoms, are related with each other by the relation called *samavāya* because the atoms are not separable from the molecule which is constructed by them, but they are only distinguishable from it.

Opponent This thesis leads to infinite regress because the *samavāya*$_1$ accepted by you must also be related with the atoms by another *samavāya*$_2$ because the *samavāya*$_1$ (between the atoms and the molecule) is again not separable from the atoms, but only distinguishable from them. Again *samavāya*$_2$ which lies between the atoms and the *samavāya*$_1$ must be related with the atoms by *samavāya*$_3$ and so on upto infinity.[28]

(II) Disputant Absence of a thing is judged by the non-apprehension of the thing. Naturally we first apprehend that we cannot apprehend the thing which is otherwise apprehensible, and therefrom judge that the thing must be absent.

Opponent — But the non-apprehension of the thing is the absence of the apprehension of the thing. Thus before you apprehend the non-apprehension of the thing, you have to apprehend the non-apprehension of the apprehension of the thing. Again before that you have to apprehend the non-apprehension of the apprehension of the apprehension of the thing and so on upto infinity. Thus before you come to judge the absence of the thing you have to undergo infinite number of apprehensions (of non-apprehensions) which is unacceptable. Thus you are committing *anavasthā*.[29]

The examples cited here are of the first type (concerning infinite number of objects) and the second type (concerning infinite number of events) of *anavasthā* respectively.

§ 7.77

Sometimes it may be found that a thesis leads to some 'infinite regress' but the 'infinite regress' is not a 'vitious infinite regress' and hence does not amount to a fallacy. That is to say, though we are led to accept infinite number of entities on accepting a thesis, acceptance of those infinite number of entities does not involve any inconsistency. Let us observe the following dialogue :

Disputant — Whatever the person is compelled to do in this life, is caused by the person's actions in his previous life.

Opponent — But then, in the same manner, whatever the person did in his previous life (let us call it L - 1), is caused by his actions in the life previous to L-1 (let us call it L- 2). Whatever the person is compelled to do in L-2 is caused by what he did in L-3 and so on upto infinity. Thus it would imply that every person has infinite number of previous lives. Thus you are led to infinite regress.

Disputant — I am not committing any fallacy called infinite regress. Because though my thesis leads us to accept infinite number of previous lives of the same person, I do accept this as my another thesis, because the world according to me has no beginning. So every person does have infinite number of past lives.[29a]

Here the answer given by the disputant to the claim of *anavasthā* made by the opponent is a typical answer often given by the thinkers who believe in the doctrine of *karma*. The above example provides us a way in which the scope of *anavasthā* as a fallacy should be restricted.

The necessary condition for any *anavasthā* to be a fallacy seems to be that the proposition which proposes infinite number of objects/events of such an such a kind to be there in the world, should be really an unacceptable proposition. The proposition does not become unacceptable just because it asserts the existence of infinite number of objects/events of such and such kind in the world, but we should have some other grounds to demonstrate the unacceptability.

Jayanta points out in *NMr* that *anavasthā* can function as a fallacy only if it destroys the thesis, that leads to *anavasthā*[30] Thus even though a debater shows that if the thesis t is accepted, we have to accept infinite number of objects/events of such and such character, the other debater can ask, 'So what? How does this infinite regress destroy t ?

While Jayanta is talking of the destruction of a thesis *(mūlakṣaya)*, he is not concerned with acceptability or unacceptability of *anavasthā* to which the thesis leads us, but he is concerned with the question, whether in spite of its truth (acceptability) the *anavasthā* makes it impossible at the same time to hold the thesis.[30]

Thus in the first example of *anavasthā* given in § 7.76, the disputant can ask, "Let there be infinite number of *samavāyas*. How does it affect the thesis that there is *samavāya* between the atoms and the molecule?" But in the case of the second example in § 7.76, a parallel question raised by the disputant will be ineffective according to Jayanta because the infinite number of previous apprehensions (in this life) will practically make the knowledge of absence impossible. Thus here *anavasthā* does destroy the thesis.

Now the above condition of *anavasthā's* being a fallacy, as laid down by Jayanta is more restrictive as to the scope of *anavasthā* as a fallacy, than the one stated by me. (The condition stated by me is identical with the one accepted by *Navyanyāya* logicians.) I suppose that Jayanta's condition of *anvasthā* as a fallacy is over-restrictive because even in the case of the first example in § 7.76, the opponent can ask a significant question, "Is such a situation, in which there is an unending chain of infinite number of relations between the atoms and the molecule, an acceptable situation?" If the situation is really unacceptable then we can conclude that the thesis namely there must be *samavāya* between the atoms and the molecules must also be unacceptable. This follows by the rule of modus tollens, because the unacceptable situation is implied by the thesis.

§ 7.78

In addition to the examples of the four types of *prasaṅga*-fallacies discussed in §§ 7.73-77 we may find various other examples of *prasaṅga*-fallacies which do not fall under the above four types. For instance, let us observe the following dialogue.

Disputant — The colour of a thing is preceived by eyes through the contact called *saṁyukta-samavāya*. Because then the colour is related with the thing by *samavāya* and the thing is related with the eyes by *saṁyoga*. Thus the sufficient condition of the perception of qualities through eyes, is the contact between the quality and the eyes, called *saṁyukta-samavāya*.

Opponent — If your thesis is granted then it would follow that the taste of food can be perceived through eyes. Because the food can be connected with eyes by *saṁyoga* and the relation between the food and its taste is *samavāya*. Thus the taste of food can be connected with the eyes by *saṁyukta-samavāya*.

> But the proposition that the taste of food is perceptible by eyes is unacceptable. Thus you are commiting a *prasaṅga*-fallacy.[32]

It is obvious that the above example cannot be brought under any of the first four *prasaṅga*-fallacies. The unacceptable proposition of this fifth kind of *prasaṅga*-fallacy, may be a logically impossible proposition (which is a contradictory propostion) or an empirically impossible proposition (except those falling under any of the first four types.)

§ 7.79

A variety of *prasaṅga*-fallacies which we find often used for refutations in the ancient Indian philosophical discussions is 'contextual *prasaṅga*-fallacy.' An example of this can be regarded as fallacious in a particular debate-context, but it need not be regarded so context-freely. Let us observe the following dialogue-forms.

(I) Disputant The propositon p is true, therefore q is true.

Opponent But if q, then u. And as you belong to the school of thought namely A, you cannot accept u. Because u is inconsistent with the doctrines of A. Thus u is unacceptable for you. Accordingly you have to reject q also.

(II)

Disputant p is true, therefore q is true.

Opponet But if q, then u. And you have already accepted the falsehood of u. Therefore you cannot consistently hold q.

Here we see that u is unacceptable to the disputant, but it is not 'categorically unacceptable,' when in the case of the first five types of *prasaṅga*-fallacies that we discussed in §§ 7.73 - 78, u is 'categorically unacceptable.'

Part V : Nayābhāsa
§ 7.80

When we talk of logical fallacies, generally we talk of unsound arguments. But many times the word fallacy is used in the context where there is no explicit argument made by the disputant which can be claimed to be sound or otherwise. However, when a philosopher is formulating certain definitions, or giving descriptions or explanations, he may be guided by certain false presuppositions which serve as the implicit premises of his conclusions. The other philosopher trying to discover a fallacy in that definition, description or explanation, points out the falsity of the presuppositions by making explicit the hidden premises of his conclusions. In this sense we can talk of the fallacies of definition, fallacies of description or fallacies of explanation.

Nayābhāsas that are put forth by Jaina logicians, may be classified as the the fallacies of description. We come across many descriptive sentences in our ordinary language. The sentences may be used in literary sense or in derivative sense. The sentences may be true when we consider some sense of it and false when we consider some other sense of it. The correct sense is determined by the context in which the sentence is used. However the philosophers who are misled by ordinary language are engaged in the context-free interpretations of such sentences and implications of such context-free propositions meant by the sentences. The expressions in ordinary language generally mislead the philosophers, while common men are satisfied with the contextual meaning expressed and the pragmatic purpose served by them.[33] That is why all the examples of *nayābhāsa* given by the Jainas are the various philosophical positions held by the philosophers of the various schools (except Jainism itself).

It is like this. According to Jainas every thing has infinite number of properties. *(anantadharmātmaka)* However all of these properties are not physical properties, but most of them are the plausible impositions of language. As our ordinary language is ambiguous, we can make seemingly opposite descriptions of one and the same thing and relate many descriptions with a thing which are not directly concerned with that thing. Thus a yellow flower can be described as red 'in the context of' some other flower. In other words, we can say, 'supposing we dub this yellow flower as some red thing we can describe it as a red flower' and so on.

Thus including counterfactual, relational, figurative descriptions (which are all contextual descriptions) we can in principle describe one and the same thing in infinite number of ways. Now Jainas argue that to regard only one sort of description as the genuine sort and to over-generalise it is to commit *nayābhāsa.* We have to regard according to them even seemingly contradictory descriptions as genuine descriptions keeping in our mind that the opposition is superficial because the contexts of the two descriptions are different.

Nayābhāsa means pseudo-*naya*, i.e. something which appears as *naya* but is not *naya. Naya* is the contextual (and thus relative to the context) truth conveyed by an ordinary language sentence.[34] A philosopher jumps from such ordinary language statement to a philosophical statement which is supposed to convey some context-free and in this sense absolute position. Such a position looks like a truth because the philosophical sentence also is a sentence. But it is not a truth because while conveying any such seemingly absolute truth, the philosopher makes overgeneralisation of the relative (contextual) rules and many a time commits category-mistakes.

§ 7.81

Jainas seem to hold that the whole of our ordinary language discourse is made up of ambiguous sentences. The sentences are true or false in relation to the senses attached to them. Again, according to Jaina view of language we have only to be aware of this fact about the language. We have only to see how the same sentence is true in one context and false in another. Their theory of *syāt* (='may be,' 'possibly,' 'in a sense,' 'in a context') points to the same attitude. The result is that Jaina philosophers as a matter of fact have not seriously tackled the problem of removing ambiguities. On the contrary they seem to try to frustrate any efforts towards precision, by pointing out or if it is not possible, by imposing the ambiguity in what the other debater says. Supposing that a Cārvāka philosopher argues that human self is moral, an orthodox Jaina would say, for instance, that in a certain sense (X) the human self is mortal, while in some other sense (Y) the human self is immoratal. He will thus pretend to satisfy Cārvāka and rest there. But has he answered the philosophical position of Cārvāka? No, because the Cārvāka's actual position is that the human self is mortal even in the other sense (Y). Jaina philosopher has to show, if he can, that human self is definitely immortal in the other sense (Y). But the Jaina only states his position and does not prove it. Thus it seems that though the other philosophers may have a faulty view towards language, Jaina view towards language also is faulty becuase they do not seem to see that up to a point the contexts and meanings can be unambiguously fixed, and after we fix them, a definitive universal view can be held and argued for, without committing any *nayābhāsa*.

§ 7.82

Though Jaina philosophers are in a sense ordinary language philosophers, their analysis of the ordinary language descriptions as well as of philosophical descriptions is prejudiced by their metaphysical presuppositions. That is why they tend to claim that all other philosophers have committed *nayābhāsas*. Jaina logicians do not give any detailed explanation why a particular philosophical doctrine is defective due to *nayābhāsa*. On many such occasions of metaphysical enquiry Jainas are found to be involved in making serious definitive claims about the reality. These claims are again philosophical claims assuming a fixed context and unambiguity of language. (For instance the claims that *jīva* is of the same extension as body, that *jīva* is never annihilated, though it may undergo change, and so on). The claims of Jainas in fact exhibit *ekāntas* (definitive results) drawn by them, though sometimes Jainas clothe them in the language of *anekānta*. Thus Jainas also commit *nayābhāsas* in the way as others do, when they try to make definitive claims assuming a fixed context and fixed sense. Thus many times the metaphysical enquiry of Jainas and their view of language do not go together.

§ 7.83

It seems that we have here to distinguish between material aspect of fallacies and formal aspect of fallacies.[35] We may be in a position to approve of the basic structure of *nayābhāsas* as observed by the Jaina logicians, but may not approve of their actual exemplification as presented by the Jaina philosophers. Leaving aside the metaphysical presuppositions behind the Jainas' account of *nayābhāsa* which are reflected in their exemplification of *nayābhāsas* we can generally say that when a philosopher is misled by ordinary language expressions and makes definitive descriptions of the reality, he is committing the fallacy called *nayābhāsa*.

§ 7.84

Though the Jaina concept of *naya* is as old as the canonical literature of Jainas, the concept of *nayābhāsa* seems to have appeared later. Umāsvāti is seen to explain the term *durnaya* in his *Panāśaka*.[36] Samantabhadra also is seen to distinguish between factual *(vastu) naya* and pseudo *(mithyā) naya*[37] However, an attempt in the classification of *nayābhāsas* corresponding to the classification of *nayas* seems to be first made by Akalanka in his *Laghīyastraya*.[38] Akalaṅka's attempt was the primary step where he could not give the account of all the seven *nayābhāsas* and gave the account of only the first three. The account of all the seven nayābhāsas is clearly found first in Vādideva's aphorismic treatise *Pramāṇanayatattvālokālaṅkāra*.39 In the following sections (§§ 7.85-91) we shall consider one by one the seven *nayābhāsas* and their examples.

§ 7.85

Naigamābhāsa

Sometimes we find that the two objects which are not separable from each other, though distinguishable, are explained by a philosopher in such a manner that they would appear as separate entities.[40] For instance thing and its qualities, thing and its movement are sometimes described as if they are separate. Ontologically thing and its qualities cannot be separated from each other. The confusion occurs because in ordinary language we have separate words for things and qualities. Thus the philosophers who are victims of the systematically misleading structure of our language may tend to think that the two distinguishabilities, namely the distinguishability (which is inbuilt in separability) between one thing and the other and the distinguishability between thing and its qualities, are on par. Jainas would say that if a thing and its qualities are regarded as entirely distinct by a philosopher, then the philosopher is committing *naigamābhāsa*, a variety of *nayābhāsa*. Jainas would say that though in a sense a thing is distinct from its qualities, in some other sense a thing and its qualities are one and the same thing.

Jaina logicians claim that *Nyāya-Vaiśeṣika* philosophers commit this fallacy. Though there is some sense in Jainas claim that Nyāya-Vaiśeṣika philosophers commit *naigamābhāsa* because the Nyāya-Vaiśeṣika philosophers regard substance and its qualities as entirely distinct reals, which can never be 'one and the same thing,' still Jainas also neglect that Nyāya-Vaiśeṣika philosophers also accept that substance and its qualities cannot be separated from each other.

'*Naigama*' is somtimes derived from *nigama* which means purpose or plan *(saṅkalpa)*. Sometimes in our ordinary language the planned or proposed actions are described as if they are presently being done. This is a kind of temporal imposition. A person carrying fuel, water and rice, when asked 'What are you doing?' says 'I am cooking', when cooking is his proposed action and not the present action. However, a thinker may be misled by such events and may conclude that the future action is implicity there when it is proposed. Such a theory would be then a case of *naigamābhāsa*. Accordingly the doctine of *satkārya* (according to which the effect is implicity present in the cause) is blamed by Jainas as a case of *naigamābhāsa*

§ 7.86

Saṅgrahābhāsa

Jaina logicians distinguish between *parasaṅgraha* and *aparasaṅgraha* as the two types of *saṅgraha*. By *parasaṅgraha they mean the view by which all the things in the world are regarded as one. By aparasaṅgraha* they mean the view by which some universal (but not 'the' universal) aspect of the things is emphasised and the peculiarities are neglected. In both the types of *saṅgraha* the peculiarities are neglected thought not denied.

In our language we can describe all the things in the world as real. However, while describing the things in this manner we may not deny that the things, in spite of having the common feature called realness, are still distinguishable from one another. But a Vedāntin may get perturbed by such an obviously true description and start arguing that the common feature, namely realness of all the things, is the reality and everything else is just an appearance. Thus the common-sense statement 'Everything in the world is real' is the expression of the *naya* called *parasaṅgraha*. But a philosopher's statement, 'Everything in the world is nothing but real, and it is this Real which is called Brahman' is an expression of *parasaṅgrahābhāsa*[41]

Similarly we may occasionally make a statement such as 'What I need is some table. It does not matter if it is made of wood or iron or anything else.' In such a statement we emphasise the common property namely tableness, but we do not deny the peculiarities of each table, though we neglect them. But a platonist may be moved by such a common-sense statement and may be led to think that tableness is real but the peculiarities of the tables are illusory. In that case the platonist is committing a *nayābhāsa* called *aparasaṅgrahābhāsa*.

§ 7.87

Vyavahārābāsa

When we do *saṅgrahanaya*, we conceive of the compilation *(saṅgraha)* of various objects to form one or more groups according to the common properties amongst the objects concerned, and treat the members of any group as non-distinct from each other. Though we may otherwise accept the peculiarities of the things, we neglect them while doing *saṅgrahanaya*. On the other hand while doing *vyavahāranaya* we concentrate only upon the peculiarities of the objects and neglect the common properties amongst them. A philosopher who is moved by such statements which emphasise the peculiarities, may therefrom conclude that only peculiarities are real whereas the commonness is apparent and unreal. Such a view would be called as *vyavhārābhāsa* by the Jaina logicians. Thus the view that only *sāmānyas* are real and *viśeṣas* are unreal would be a fallacious view due to *saṅgrahābhāsa*, and the view that only *viśeṣas* are real and *sāmānyas* are unreal would be fallacious due to *vyavahārābhāsa* according to Jainas.

It may be seen that some Buddhist philosophers believed that there are bare particulars *(svalakṣaṇa)*, there is nothing common between any two particulars and the similarity amongst them is imposed by mind. This view is similar to the 'fallacious view' termed as *vyavahārābhāsa*. However, this Buddhist view does not seem to be the result of the misleading nature of language.

Akalaṅka understands by *vyavahāranaya*, the commonsense view in which the existence of the external objects is accepted. According to him, the philosophers like Śūnyavādin, Vijñānavādin who deny the existence of all objects or external objects commit *vyavahārābhāsa*.[42] It should be noted here again that this type of *vyavahārābhāsa* does not seem to be caused by the misleading nature of our day-to-day language. It is possible, however that a philosopher may be misled by the ordinary language expression like 'When I was talking of the lecturer in philosophy, I had you in my mind.' Such sentences may mislead a thinker to think that if anything can be there in one's mind, why not all the possible things? But generally such a thesis of the denial of external world arises from (a correct or incorrect) analysis of human knowledge, and not from a correct or incorrect analysis of human language.

Sometimes *vyavahāranaya* is regarded as the view by which the figurative use of language is taken to be derivatively meaningful.[43]

Thus to regard such expressions to be literally meaningful would amount to committing *vyavahārābhāsa*. (This is somewhat like *upacāracchala*. See § 7.4) Therefore any serious attitude to hypostatise the concepts, personify the unconscious objects and so on, will be called as committing *vyavahārābhāsa*.

Curiously enough Vādideva suggests that the view of Cārvāka is fallacious due to *vyavahārābhāsa*.[44] While explaining *vyavahāranaya* he suggests that holding the separability between soul *(jīva)* and body, distinguishing between different grades of soul etc. is the task which is involved in doing *vyavahāranaya*. Cārvāka distinguishes between different elements of matter (namely the four gross-elements) and also holds that they are separable from each other. But he does not accept the separability between soul and body, he does not approve of different grades of souls. Thus, according to Vādideva, Cārvāka commits the fallacy of *vyavahārābhāsa*. It is clear that here Vādideva is employing the Jaina methaphysical dogmas for settling the logical issue.

§ 7.88

Ṛjusūtrābhāsa

When we do *ṛjusūtranaya*, we are concerned only with the object which is given to us in immediate present and we neglect the past and the future objects and even the present ones of which we are not directly aware. We do not consider, though we do not deny that the same object had some modes in the past and would have some modes in future. We may describe such an object by using the expressions as 'this red' or 'here now yellow' or 'Here is now the feeling of pleasure.'[45] A philosopher may get inspired by such an attitude and comment that the descriptions like 'this red,' 'here now yellow' etc., are the only genuine descriptions of reality. Only the present momentary object of one's knowledge is real and anything else is merely a mental construct. Jaina logicians regard such a view as fallacious due to *ṛjusūtrābhāsa*. Jainas also claim that Buddhists commit this fallacy when they hold that all the objects are the momentary sense-data and continuity of the so-called substance is illusory.

§ 7.89

Śabdanayābhāsa

Tense of the sentence, the case-endings of the words, the genders of the nouns, singularity or plurality and suffixes of the word-forms are the factors which are responsible for the difference in meaning of a sentence. One who is doing *śabdanaya* emphasises these factors. However, a thinker may over-emphasise these factors to be responsible for the changes in the word-meanings. Take for instance the following cases.

(i) Take the two statements,

1. There was a mountain called Meru.

2. There is a mountain called Meru.

The statements certainly differ in meaning. But a thinker may further argue that because the meaning of the statement changes, accordingly the meaning of the word Meru also should change. This fallacy would be called *Śabdanayābhāsa.*

(ii) Consider the two statements,

1. The teacher taught the subject,

2. The subject was taught by the teacher.

In these sentences the positions of 'teacher' and 'subject' are changed. But the meanings are not changed. However, to argue that the meanings have also changed because the positions have changed would be to commit *Śabdanayābhāsa.*

(iii) In Sanskrit the change in gender does not necessarily mean the change in meaning. The words *taṭa* (masculine), *taṭī* (faminine) and *taṭa* (neutral) mean the same thing namely bank (of a river, pool etc.). Similarly the words *dāra* (masculine plural), *kalatra* (neutral) and *patnī* (feminine) are synonymous and mean 'wife.' But a thinker may speculate that because the genders are different, the meanings must be different. Such a speculation would be fallacious due to *śabdanayābhāsa.*

(iv) In Sanskrit the word *jana* (= people) is used either in singular form or in plural form. But the meaning of the word does not differ due to singularity or plurality. But a thinker who argues that the meaning must differ because the number differs, is committing *śabdanayābhāsa*.

(v) The Sanskrit verbs change their meanings according to the prefixes attached to them. (उपसर्गेण धात्वर्थो बलादन्यत्र नीयते ।) Though this is generally true, sometimes the prefixes remain neutral and do not change the meaning of the verb. For instance we find that the terms *jñāna, vijñāna, prajñāna* are used in Upanisads synonymously. However, one who argues that the meaning must change according to the prefix, is committing *śabdanayābhāsa*.

In all the above cases the person who commits *śabdanayābhāsa* makes over-generalisations of the rules of language. Though the game of language needs rules, the rules of language need not be hard and fast, and there can always be some exceptions to most of the rules. The philosopher who commits *śabdanayābhāsa* denies this possibility. He conceives of language as a perfect mechanical system.

Sometimes *śabdanaya* is taken to mean the view by which the role of words in making the things known is emphasised. Accordingly *śabdanayābhāsa* is supposed to mean over-emphasis on the role of words. Thus one who claims that knowledge without the mediation of words is impossible, commits *śabdanayābhāsa*. Thus the philosophers of Vyākaraṇa-school who hold this view are supposed to commit this fallacy.[46]

§ 7.90

Samabhirūḍhābhāsa

The person who is doing *samabhirūḍhanaya* emphasises that the meaning or shade of the meaning of a word differs according to the etymology of the word. It is true that in our language we may use a particular word avoiding the other purposely, to express the shade of meaning that is expressed by its etymology. But this cannot always be the case. The two words which differ from each other etymologically may be used as the exact synonyms of each other. For instance, the three Sanskrit synonyms *Indra, Śakra* and *Purandara* are supposed to denote the some object namely the King of gods. But the etymological meaning differs. For instance *'Indra'* is derived from the root *ind* (= to be prosperous) and therefore etymologically means 'prosperous'. *'Śakra'* is derived from the root *śak* (= to be powerful) and therefore etymologically means 'powerful'. And 'Purandara' is derived from *'pūr'* (= city) and *'dṛ'* (= destroy) and therefore etymologically means one who destroys the cities. Thus the different synonyms have different connotations. But a thinker may wrongly argue that because the etymologies are different, denotations also must be different. Such a view is an example of *samabhirūḍhābhāsa.* (The synonymous terms 'morning-star' and 'evening-star' also can be treated in the similar manner.) This *nayābhāsa* means the over-emphasis on etymological meaning (and the connotation suggested by it) and the confusion between connotation and denotation.

§ 7.91

Evambhūtābhāsa

We name the things and create conventions. While giving a particular name to a thing we may take into consideration the function served by the thing. But after the thing is so named, the name becomes a conventional symbol to connote that thing, whether or not the thing continues to serve the particular function. For instance, a person is called a lecturer if he delivers lectures. But we could call that person a lecturer even when, for instance, he is asleep. We may say, 'The lecturer is sleeping.'

But a philosopher may think that the 'functional' name should be used only when the named object serves the function. A person should be called a lecturer only when he is lecturing and not otherwise. The fallacy in this view according to Jainas is *evambhūtābhāsa.*

This fallacy can be treated as a species of *samabhirūḍhābhāsa.* Because the thinker who expects that a person should be called a lecturer only when he is lecturing, confuses between the etymological meaning of 'lecturer' (which is the connotation) and the conventional reference of 'lecturer' (which is the denotation). But the difference between the two is that, (1) in *evambhūtābhāsa*, the function of the denoted object is considered and the structure of it is neglected, while in *samabhirūḍhābhāsa*, both structure and function are considered; and (2) in *evambhūtābhāsa* time-element (that is, the time of serving the function and the time of calling the thing by the name concerned should be one) is important, while in *samabhirūḍhābhāsa* it is not so.

§ 7.92

Though the above classification of *naya* and *nayābhāsa* does not appear to be precise and exhaustive, it can still help us in understanding different forms of misleading expressions. It may also be seen that Jainas here are also defending the common-sense view of the world against the so-called philosophical views presented by various philosophers (of the schools other than Jainism). Some such common-sense views can be cited as follows :

1. Thing and its qualities cannot be separated.

2. Time is not one. Past, present and future are distinct.

3. There are particulars and the plurality of things is true.

4. There are universals (common properties, classes).

5. External objects exist.

6. There are substances which continue to exist in the passage of time.

7. The language is not a mechanical system.

8. The meanings of the words are not fixed merely by etymology; the convention plays a very significant role in deciding the meanings of words.

Out of these *naigamābhāsa* is against (1), (2). *saṅgrahābhāsa* is against (3), *vyavahārābhāsa* is against (4), (5). *ṛjusūtrābhāsa* is against (6). *Śabdanayābhāsa* is against (7). *Samabhirūḍhābhāsa* and *evambhūtābhāsa* are against (8).

Though Jainas are seen here to defend the common-sense view of the world, they cannot be called common-sense philosophers in the sense in which G.E. Moore is one. Moore would not accept that the genuine common-sense beliefs are in a way defective. Moore, with all his efforts would justify the common-sense beliefs and try to analyse them in the way they could be clearly understood.[47] But Jainas would not always justify the common-sense beliefs. They would suggest that though in a sense (or in a way, from a certain point of view) these beliefs are true, is some other sense they are false. These beliefs are partial (inexhaustive) views of the things and in this sense defective visions of reality. To claim that they are utterly false is to commit *nayābhāsa*. But even to claim that they are not at all defective also is to commit *nayābhāsa*. Thus Jainas' defence of common-sense is conditional and half-hearted while Moore's defence of common-sense is categorical and well- thought out.[47a]

PART VI : Fallacies of Definition

§ 7.93

Fallacies of definition are generally the fallacies of false inductive reasoning, because definitions as conceived of in the Indian tradition are the results of inductive generalisation. When X is defined as Y, it is claimed that all cases of X are necessarily the cases of Y and vice versa. (But it is not claimed that X 'means' Y in which case the definition would be an analytic statement.) Thus according to Indian logicians there should be the relation like material equivalence between the definiendum (= the term defined) and the definiens (= the term in which definiendum is defined. In other words if D is the definiendum and D' is the definiens, then D' must be a distinctive featue of D and there must be material equivalence between Dx and D'x for all values of x. (Dx means x has the property D-ness; D'x means x has the property D'. For instance if earth is definiendum and odour is the definiens then the material equivalence means that for every x if x is earth then x has odour and vice-versa.)

Accordingly Nyāya logicians conceived of three fallacies of definition namely *avyāpti*(too narrow a definition)*ativyāpti* (too wide a definition) and *asambhava* (impossible definition).

One who tries to claim the fal!acy called *avyāpti* in a definition, tries to show that definiendum does not apply to all the cases to which definiens does. In other words there is at lest one case in which D holds good but D' does not. For instance if cow is defined as a black animal then it is fallacious on account of *avyāpti* because we can point out at east one cow (D) which is not a black animal (D')

One who tries to claim the fallacy called *ativyāpti*, tries to show that definiens does not apply to all cases to which definiendum does. In other words there is at least one case in which D' holds good but D does not. For instance if cow is defined as an animal having horn then the definition is fallacious on account of *ativyāpti* because we can point out at least one animal having horn which is not a cow.

One who tries to claim the fallacy called *asambhava*, tries to show that definiendum applies to only those cases to which definiens does not apply. For instance if cow is defined as an animal having only one horn then the definition will be fallacious on account of *asambhava* because there is no cow who has only one horn.[47b]

§ 7.94

Though in the above list of the fallacies of definition we do not find the mention of mutual dependence *(itaretarāśraya)*, the philosophers like Jayarāśi are seen to opine that mutual dependence also is a fallacy of definition. It is the fallacy of epistemic mutual dependence. This is as follows. The intension in defining a term is making the term known clearly and distinctly. So if any definition is to be pragmatically significant, at the moment when it is presented, the definiens must be already known clearly and distinctly. The function of the definition then is to bring us from the clearly and distinctly known definiens to the definiendum not known clearly and distinctly, and thus to make the definiendum known by pointing out the definitional equivalence between the two. But if we need the clearcut knowledge of the definiendum itself for knowing the definiens clearly and distinctly, then there would occur the fallacy of mutual dependence. Jayarāśi in his *TUS*, tries to point out that any attempt in defining the true cognition does involve this fallacy.[48]

Notes

1. See §1.2
2. It seems that at the time of Akṣapāda *upacāracchala* was generally included in *vākchala*. For instance, in *upāyahṛdaya* ascribed to Nāgārjuna, there is an example of *chala* in which the sentence 'The mountain is burning' meaning the trees and the grass on the mountain are burning) is misinterpreted as 'The rocks etc. of the mountain are burning.' Though this example quite suits to *upacāracchala*, it is described as an example of *vākchala* by the author of *Upāyahṛdaya*. (See *Upāyahṛdaya* p. 15 in *PBTL*)

2a. I owe the use of the words disputant and opponent for *vādin* and *prativādin* respectively, to VIDYĀBHŪṢAṆA. Of course, I also feel that *vādin* and *prativādin* are relative terms and can be interchangeably applied to the two debaters involved in a debate. It may be noted that I have not considered the relation between disputant and opponent as parallel to one between *pūrvapakṣī* and *siddhāntī*.

3. See § 7.?9
4. See for instance § 7.7
5. According to the author of *Upāyahṛdaya*, it seems, *jāti* does not mean necessarily a fallacious refutation because he terms *jātis* only as the refutations *(dūṣaṇa)*.
6. See *NVT*, Pt. II, p. 674.
7. विशेषहेत्वभावाच्च अनैकान्तिकदेशनाभासा इति । *NV*, p. 531
8. See *BTR*, p. 201.
9. See TUCCI's comparative list in *PBTL*, Intro., p. xxi.
10. See *NM* pp. 54-71. Also see *HL*, p. 70 (Note that Socmock which is referred to in *HL* is nothing but Akṣapāda. See *HL* p. 21, fn. 3).
11. छलव्यवहारोऽपि विजिगीषूणां वाद इति चेत् न ।*VN* p.68. Also, तत्त्वरक्षणार्थं सद्भिः उपहर्तव्यमेव छलादि विजिगीषुभिरिति चेत् ...... तस्मान्न ज्यायान् अयं तत्त्वरक्षणोपायः ।VN p. 69, Compare this with *NS* 4.2.50.
12. साधनप्रख्यापनं सतां तत्त्वरक्षणोपायः साधनाभासदूषणं च । *VN* p. 69
13. My account of the accasions of defeat here is based on *NS* 5.2 and *NBh* under it.

14. See § 2.14.

15. See *NBh* on *NS* 5. 2 . 9. Caraka had prescribed such tricks to the debaters (who are medical practitioners). See § 1.3.

16. The sixteen terms enlisted by Akṣapāda are as follows.

    1) *pramāṇa*, 2) *prameya*, 3) *saṁśaya*, 4) *prayojana*, 5) *dṛṣṭānta*, 6) *siddhānta*, 7) *avayava*, 8) *tarka*, 9) *nirṇaya*, 10) *vāda*, 11) *jalpa*, 12) *vitaṇḍā*, 13) *hetvābhāsa*, 14) *chala*, 15) *jāti*, 16) nigrahasthāna (See *NS* 1.1.1)

17. The respective verse in *VN* runs,

    असाधनाङ्गवचनम् अदोषोद्भावनं द्वयोः ।

    निग्रहस्यास्थानमन्यत्तु न युक्तमिति नेष्यते ।। *VN* pp. 4-5.

18. तस्यैव साधनस्य यत् नाङ्गं प्रतिज्ञोपनयनिगमनादि तस्य असाधनस्य साधनवाक्ये उपादानं वादिनो निग्रहस्थानं व्यर्थाभिधानात् ।
    VN p.59.

    Also,

    न प्रतिज्ञादीनां न्यूनं हीनं तदभावे प्रतीतिभावात् । *VN* p.104.

19. For instance, Diṅnāga sees it. See § 3.1.

20. अत्रापि न कश्चित् क्रमनियमः इष्टार्थसिद्धेरुभयत्र अविशेषात् ।

    VN, p. 8. Also see *VN* pp. 98-104.

21. यश्चोभयोः समो दोषः परिहारस्तयोः समः ।

    नैकः पर्यनुयोक्तव्यः तादृगर्थविचारणे ।।
    general rule (*'nyāya'*). See for instance, *SDS*, p. 438.

22. For instance *'prasaṅgasama'* of Akṣapāda suggests his awareness of the fallacy of infinite regress (= *anavasthāprasaṅga*). See § 7.17. Also see § 7.32.

23. *'kāraṇopapattitaḥ'*. See *NS* 1.1.40. Vātsyāyana's interpretation of *kāraṇa* as *pramāṇa* and *hetu* (See *NBh* thereon) seems to be artificial.

24. Jayanta for instance argues that the proof by *prasaṅga*-method, as such is not possible. He says,

    प्रसङ्गसाधनं नाम नास्त्येव परमार्थतः ।

    तद्धि कुडयं विना तत्र चित्रकर्मेव लक्ष्यते ।।
    NMr, Pt. I, p.95.

25. The dialogue is based on Nāgārjuna's discussion on possibility of *pramāṇa*, that occurs in *Vigrahavyāvartanī*, verses 47-50. (See Appendix 5 of *MS*).

27. The dialogue is based on Cakrapāṇidatta's commentary on the first sentence of *Carakasamhitā*. See *CS (1922), pp. 1-2.*

28. The dialogue is based on Śaṅkara's discussion of *samavāya* in *BSSB* 2.2.12.

29. The dialogue is based on Jayanta's discussion of *abhāva*, where Buddhists are said to claim that one commits the fallacy called *anavasthā* when one's notion of *abhāva* is that of a negative entity. Buddhists avoid this fallacy by defining *abhāva in positive terms*. See *NMr*, Pt. I, p. 54.

29a. Vācaspati advances a similar argument for the beginningless series of the false suppositions *(adhyāsa)* and objectivity of objects *(viṣayatva)* when he says,

'न चाध्यासे सति विषयत्वं विषयत्वे चाध्यास इत्यन्योन्याश्रयत्वम् इति साम्प्रतं, बीजाङ्कुरवदनादित्वात् ।', *BHAM*, p.23.

Though the argument is advanced against the claim of 'mutual dependence,' actually it should be taken as against the claim of *anavasthā* because the opponent's claim is not that the same occurrences of *adhyāsa* and *viṣayatva* depend on each other but that every occurence of *adhyāsa* depends on some prior occurrence of *viṣayatva* and vice versa.

30. मूलक्षतिकरीमाहुः अनवस्थां हि दूषणम् ।
मूलसिद्धौ त्वरुच्यापि नानवस्था निवार्यते ।।
NMr, Pt. I, p. 21.

31. Jayanta's example is like this. The Buddhist argues that if *pramāṇa* is used even to know the already known objects then what prevents it from knowing at every moment the same object? Thus Naiyāyika has to accept infinite number of cognitions of the same object. On that Jayanta argues that such a consequence, though may be true or false, does not make the thesis (namely *pramāṇa* can be used to know an already known object) impossible. See *NMr*, Pt. I, p.21.

32. This example is based on Jayanta's discussion of the perceptibility of *abhāva*, where he says,

यस्तु संयुक्तविशेषणभावे सन्निकर्षे रसादिभिः अतिप्रसङ्गः उद्भावितः सोऽयं समवायाख्ये चक्षूरूपसन्निकर्षेऽपि समानो दोषः । *NMr*, Pt. I, p. 49.

33. Such a view is clearly held in the modern philosophy of west by RYLE. See his paper 'Systematically misleading expressions' published in *PAS* Vol. XXXII (1931-32), pp. 139-70.

34. *Nayavāda* seems to be an appropriate term for ordinary language expression which is context-bound. For instance, Sidhasena Divākara, an ancient Jaina logician, says

जावइया वयणवहा तावइया चेव होन्ति नयवाया । *STS*, p. 284.

(Meaning : There are as many *nayavādas* as there are ways of speaking).

35. I have made this distinction somewhat on similar lines to BARLINGAY's distinction between formal mode and material mode of *pramāṇa*. See *MIIL*, p. 17.

36. चेदेकांशकलङ्कपङ्ककलुषास्ते स्युस्तदा दुर्नयाः ।
Pancāśaka by Umāsvāti. (Referred to by DEVENDRAMUNISHASTRI, *JSV* P. 320, fn 2 *(Kha)*).

37. निरपेक्षा नया मिथ्या सापेक्षा वस्तु तेऽर्थकृत् ।
Āptamīmāṁsā, verse 108 (included in *AMPP)*

38. See *LT* pp. 57 - 61

39. See *PNTA*, Chapter VII.

40. The distinction between distinguishability and separability is made here following BARLINGAY's article entitled 'Distinguishables and Separables' (IPQ, Vol. II, No. 2, Jan 1975).

41. Akalaṅka states this when he says ब्रह्मवादस्तदाभासः ।*LT*, p. 57.
In Western philosophy GALLIE refers to a similar fallacy called essentialist fallacy. See his article 'The function of philosophical Aesthetic' (pp. 13-14) published in *Aesthetics and Language.*

42. See *LT*, p. 61.

43. See *JMM*, p. 359. But SANGHAVI brings this type under *naigama. See what he says in TAS*, p. 58.

44 See *PNTA*, pp. 524-525.

45. सुखविवर्तः सम्प्रत्यस्ति ।*PNTA* 7.29.

46. न सो ऽस्ति प्रत्ययो लोके यः शब्दानुगमाद् ऋते ।
अनुविद्धमिव ज्ञानं सर्वं शब्देन भासते ।।
VPB, verse 123. Also see JMM, pp. 371-2.

47. See MOORE's article entitled 'A defence of Common Sense' published in *PP*, pp. 32-59.

47a. This part V of this chapter was published in Saṁbhāṣā (Nagoya University, Japan), Vol. 11 in the form of an article entitled "The Jaina doctrine of Nayābhāsa". I thank the editors of the journal for the same.

47b. Symbolically this can be presented as follows. A correct definition should satifsy the equivlence condition which states $(x)\ (Dx \equiv D'x)$. The fallacy of definition viz. *avyāpti* occurs when the situation of the form $(\exists\ x)\ (Dx.{\sim}D'x)$ is found which amounts to violation of the equivalence-condition. *Ativyāpti* occurs when the situation of the form $(\ \exists\ x)\ (D'x.{\sim}Dx)$ is found. *Asambhava* occurs when $(x)\ (Dx \supset {\sim}D'x)$ is found to hold good.

48. I have elaborated this point in my paper entitled 'The Philosophical Position of Jayarāśibhaṭṭa,' published in IPQ, Vol. V, No. iii (April 1978).

Chapter VIII

Inference And Fallacies Discussed In Ancient Indian Logic And Western Logic : A Comparative Survey

Part I : Inference

§ 8.1

While discussing the nature of *anumāna* in Indian logic *anumāna* is translated as or identified with 'inference.' But while making such a translation or identification some basic differences between *anumāna* and inference should not be undermined.

Firstly the term *anumāna* in Indian logic is generally used in the sense of a *pramāṇa*, which implies that *anumāna* is supposed to be an authoritative means of (acquisition or justification of) a true cognition. Strictly speaking, therefore *anumāna* cannot be classified into sound and unsound, but it would be more appropriate to call the so- called unsound inference as *anumānābhāsa* (Pseudo- *anumāna) rather than unsound anumāna.* On the other hand it makes a perfectly good sense to classify inference into sound and unsound. If, therefore, for the sake of convenience we talk of sound and unsound as the two kinds of *anumāna*, we are talking in some secondary and loose sense of the term *anumāna* and not in its primary and strict sense.

Secondly, although we may take some liberty with the term *anumāna* so as to classify *anumāna* into sound and unsound, it is still different from inference. For *svārthānumāna*, which is a kind of *anumāna* does not necessarily involve implicit or explicit argument whereas inference, by the very force of meaning does so. *Svārthānumāna*, for instance as Diṅnāga understands it, assumes the form of an indirect intuition rather than inference. (See § 3.1). Inference as understood in the western logic, however, necessarily involves some process of reasoning or argumentation. In this sense of inferernce, *parāthānumāna* is certainly a type of inference but *svārthānumāna* is not necessarily so. It is true that if *svārthānumāna* and *parārthānumāna* are regarded as isomorphic with each other, the latter being the linguistic presentation of the former, *svārthānumāna* functions as a tacit inference. But such an understanding of *svārthānumāna* is more clearly available in Nyāya logic than in Buddhist logic.

§ 8.2

There is another difference that needs to be kept in mind. Western logic from its ancient period understands 'inference' to be basically of two kinds : inductive and deductive. This basic classification of inference is not available in this form in Indian Logic. This does not mean, however, that Indian logicians by and large were unaware of the distinct problems of inductive reasoning and deductive reasoning. But the status of inductive reasoning as a kind of *anumāna* has been quite dubious in Indian Logic. The problem of this status is much complicated. If our idea of inductive reasoning is broad enough to cover analogical reasoning then the whole ancient Indian conception of *anumāna* starting from Caraka to Udyotakara and even that in Diṅnāga can be seen to be revolving around inductive reasoning. The Nyāya conception of *anumāna* in this period tends to become clear if we take it as a notion of analogical reasoning strong enough to be called *pramāṇa*. *Anumāna* was conceived as a strongest possible analogical reasoning until the statement of universal *vyāpti* was introduced as an essential element of *parārthānumāna*. (We have observed this during the first four chapters.) In this period, side by side, a rough understanding of what may be called inductive generalisation must have been developing. Although the notion of *vyāpti* was not presented in Ancient Indian Logic until Dharmakīrti, logicians were concerned with the ways in which one can arrive at the knowledge of some universal truths such as causal laws. The method of inductive generalisation must have been employed by these ancient logicians and a self-consciousness of it in however faint form must have arisen in them. But one has to trace the sings of such a self-consciousness rather carefully. It is not our objective to do that here. But we will mention such a sign though passingly.

§ 8.3

We have referred to Caraka's definition of *anumāna* as *tarka* that depends on *yukti*. We translated it as a speculation based on the application of *hetu* (§ 1.5). Although such a translation is in tune with Caraka's explanation of the concept of *anumāna*; an alternative translation could be thought of by interpreting *yukti* differently. Caraka in another context[1] defines *yukti* as a cognition or cognitive faculty *(buddhi)* that sees the objects to be generated by the conglomeration of many causes. This cognition is said to be applicable to objects of all times. *Yukti* can be construed more generally as a cognition or cognitive faulty by which one grasps the eternal causal laws governing any given type of phenomena. The term *yukti* on the other hand literally means reasoning or application (of reason). If the Caraka's term *yukti* can be said to have both these shades of meaning, it can be interpreted as the reasoning involved in inductive generalisation pertaining to causal laws. Since inductive generalisation involves analogical reasoning (we have seen this in § 2.3.), it is possible that Caraka uses the term *yukti* in the sense of analogical reasoning also. Thus either the term *yukti* is used with two different but closely connected meanings or the same cognitive faulty viz. *yukti* is supposed to have two different but closely connected functions. This perhaps explains Caraka's inclusion of the term *yukti* in the definition of *anumāna*, (where *anumāna* stands for analogical reasoning.)

Apart from Caraka's vague but suggestive discussion of *yukti* as a method of causal generalisation we rarely find the explicit mention of inductive method as a kind of *pramāṇa* in the pre-Dharmakīrti Indian logic. The pre-Dharmakīrti awareness of the conditions of proper inductive reasoning seems to have expressed itself in the negative from, that is, by way of discovering inductive fallacies (which were primarily the fallacies of analogical reasoning.) We shall see in § 8.61 that many fallacies (i.e. the fallacies based on falsity/dubitability of the *vyāpti*-statement) were presented in these days in the form of the fallacies of analogical reasoning. Hence early approaches to inductive reasoning help us more in understanding how induction should not be made rather than how it is and should be made.

§ 8.4

On this background Dharmakīrti seems to have approached the problem of inductive generalisation in a much more positive and definitive manner. He tried to tackle the problem by taking recourse to the notion of essential nature *(svabhāva)* of things. Although he said that *vyāpti* can be determined either on the basis of cause-effect-relation or on the basis of essential nature as the determinant;[2] again he explained his notion of cause-effect-relation in terms of essential nature. On this view we can determine the cause-effect-relation between smoke and fire because the essential nature of smoke consists in the fact that smoke is essentially something produced by fire.[3] He also suggested that all the essential characteristics of anything can be fully grasped by experiencing it; the recourse to *anumāna* is needed only because of confusion.[4] This further implies that in so far as the determination of a *vyāpti*-relation amounts to determination of the essential nature of the entity or entities under consideration and the determination of the essential nature depends upon clear perceptual knowledge of the entity or entities under consideration, *vyāpti*-relation ultimately is determinable on the basis of clear perceptual knowledge of the essential nature of things.

This suggestion of Dharmakīrti might have influenced Nyāya logicians like Trilocana (the teacher of Vācaspatimiśra) and consequently Vācaspatimiśra who defined *vyāpti* as *svābhāvika sambandha* (essential relation) and regarded it to be determinable by some special kind of perception.[4]

Apart from this Buddhist influence there must have been some philosophical reasons too behind the *Nyāya* understanding of inductive generalisation as a variety of perceptual knowledge. When the knowledge of *vyāpti* was accepted as a necessary step towards inferential knowledge *(anumiti)* the knowledge of *vyāpti* itself could not be regarded as inferential because then it had to depend upon the inferential knowledge of some other *vyāpti* and so on ad infinitum. This difficulty could not be resolved by saying that *anumiti* based on the knowledge of *vyāpti* was one kind of inferential knowledge whereas the knowledge of *vyāpti* itself was another kind. Because the five-membered pattern of inference was fixed at the hands of Naiyāyikas and the process of the knowledge of *vyāpti* could not fit into it. For instance, in the case of *anumiti*, the 'instances' were necessarily excluded from the realm of *pakṣa*. In the knowledge of *vyāpti*, on the other hand, *vyāpti* had to include all the instances (either positive or negative or both) on which it was based. Thus the instances could not be excluded from the realm of *pakṣa* (i.e. the locus of the probandum; here *vyāpti* itself is the probandum).

But inferential knowledge i.e. *anumiti* cannot be authentic if the knowledge of *vyāpti* on which it is based is not equally authentic. Hence for Naiyāyikas, the knowledge of *vyāpti* has to be based on some *pramāṇa*. But this *pramāṇa* cannot be *anumāna* for the reasons given above. And *vyāpti*-knowledge is possible even without the application of other two *pramāṇas* viz. *upamāna* and *śabda*. Hence Naiyāyikas must have thought that *vyāpti* must be accessible to *pratyakṣa-pramāṇa* of some kind.

It appears from the above considerations that there must have been some strong theoretical reasons which prevented Nyāya logicians from straight-away classifying inference *(anumāna)* into inductive and deductive.

§ 8.5

Although the above theoretical reasons for Naiyāyikas' acceptence of intuition (or a special variety of perception) as the source of the knowledge of *vyāpti* are theoretically important, they are insufficient to the extent to which they do not take the possibility of there being fifth *pramāṇa* responsible for the knowledge of *vyāpti* seriously. This additional *pramāṇa* may be neither perception nor *anumāna* in the strict sense of the term, but still it could be a variety of 'inference' (other than *anumāna* proper). Jaina logicians seem to have conceived this possibility when they talked of *tarka* as a *pramāṇa*. This suggestion of the Jaina logicians is theoretically important because it implies that the process of inductive generalisation cannot proceed just by perceptions or intuitions but it also involves reasoning in some form. This reasoning may not be of the same kind as the one involved in *anumāna*. In Akalaṅka's scheme which is pioneering in the reconstruction of the Jaina theory of *pramāṇas*, *tarka* stands for inductive reasoning[5] and *anumāna* for a variety of deductive reasoning. This approach of Akalaṅka opened up a new horizon of logico- philosophical studies in inference; because it concentrates on reasoning-aspect of inductive generalisation and gives it an equal or equitable status with deductive reasoning. But this approach suggested by Akalaṅka and later Jaina logicians could not develop further in the history of Indian logic. This was partly because other systems of Indian logic were not open enough to welcome and develop this approach. The only impact of this move of the Jaina logicians on Nyāya logic seems to be that *tarka* was regarded as an auxiliary means to the knowledge of *vyāpti* by some later Nyāya logicians. But because according to Naiyāyikas *tarka* never gives rise to certainty but only to probabilistic cognition,[6] it was never given the status of a *pramāṇa* by them.

§ 8.6

To sum up, we find that Indian logicians approached the problem of inductive generalisation basically at three levels.

i) At one level inductive generalisation was approached negatively. There was more concentration on inductive fallacies than on the nature and adequate method of inductive generalisation. At this level the fallacies of inductive generalisation and the fallacies of analogical reasoning can be assimilated because analogical reasoning and inductive generalisation have some important features in common. We will compare this level with J.S. Mill's account of inductive fallacies in § § 8.48-8.55.

The Nyāya and Buddhist logicians who flourished before Dharmakīrti seem to have approached induction in this way.

ii) At the second level we have more concentration on the method of induction, but inductive generalisation is viewed more as an intuition or an extra-ordinary perception than as an inference. COHEN and NAGEL talk about a similar approach to induction in western philosophy. They point out that induction so understood has been called by W.E. Johnson intuitive induction.[7] Dharmakīrti and the Naiyāyikas who flourished after him aproached induction in this way.

iii) At the third level too there is concentration on the nature and the method of induction but here induction is viewed more as a kind of reasoning than as intuition. Induction in this form is different from perfect or complete induction. It is also different from demonstrative inductive reasoning.[8] It rather stands for probabilistic reasoning through which one tries to arrive at or support some general truth on the basis of some particular truths, in most cases observational truths. Akalaṅka and the later Jaina logicians approached induction in this way.

§ 8.7

Let us now turn to the classification of *anumāna* expressly made in ancient Nyāya and Buddhist logic. Perhaps the most ancient one is into *pūrvavat, śeṣavat* and *sāmānyatodṛṣṭa. Nyāyasūtra* refers to it. But the diversity of interpretations that this classification assumes in the commentaries of *Nyāyasūtra* indicates that the tradition of ancient Indian logic in which this classification was propounded was not clearly available to the Nyāya commentators. BARLINGAY makes a conjecture on this basis that perhaps the relevant aphorism of *Nyāyasūtra* is an interpolation and should be traced to Greek source.[9] But we also have another clue for interpreting the original classification. Although there is a diversity of interpretation amongst Nyāya commentators, there is no such great diversity amongst some sources outside the Nyāya tradition. TUCCI has tapped some such sources.[10] They are the two pre-Deinnaga Buddhist sources namely

Upāyahṛdaya and *Tsing Mu* and also Gaudapāda's commentary on *Sāṅkhyakārikā*. By considering the accounts of them collectively and synthetically we can draw a more or less uniform picture of the original classification. Let us consider the three types one by one.

(i) *Pūrvavat* : While making *anumāna* of this type we are, broadly speaking, operating with our belief that the laws of nature do not change over a period of time. As an implication of this the nature (or the essence) of things too does not change over a period of time. In simpler words, the present resembles the past (and naturally the future will resemble the present and the past). The examples of our operation with this belief may be diverse in form. One may make an inference with regard to somebody's personal identity over a period of time (the example given in *Upāyahṛdaya)* or one may infer a cause from its effect on the basis of the cause-effect-relation observed in the past (the example in Tsing Mu) or effect from cause again on the bais of past experiences (the example in Gaudapāda's commentary).

(ii) *Śeṣavat* : While making *anumāna* of this type we are not concerned with trying to assimilate our past experiences with the present one, but we are rather concerned with the relation between homogeneous whole and its parts. Here we are operating with our belief in the homogeneity of a given whole. On observation of one part of the whole, possessing a characteristic, we claim that the remaining parts of the whole too must be possessing the same characteristic. Thus having tasted one drop of sea-water, we infer that the remaining drops too must be saulty. Or knowing that one grain of rice is cooked, we infer that the remaining grains of the same lot must have been cooked. The central idea of this type of *anumāna*, if extended further leads to our operation with the belief in homogeneity of nature. It implies that the laws of nature or the natural order that is operative in one part of the universe must be operative in the other parts also.

In fact, it could be said, *pūrvavat anumāna* also presupposes homogeneity of nature. But it is mapped on the time-dimension. Whereas the homogeneity of nature which *Śeṣavat* presupposes is mapped on the space-dimension. The examples of these *anumānas* indicate that these two kinds of homogeneity are presupposed rather than asserted in the respective types of *anumāna.*

(iii) *Sāmānyatodṛṣṭa* : As I have suggested in § 2.4, the three fold classification of *anumāna* given in Nyāyasūtra could have been an expansion of the two-fold classification into *dṛṣṭa* and *sāmānyatodṛṣṭa. Pūrvavat* and *śeṣavat* can thus be subsumed under *dṛṣṭa. Dṛṣṭa anumāna* applies to those cases where the thesis is empirically verifiable in some sense. *Sāmānyatodṛṣṭa* on the other hand is applicable to those cases where the thesis is not so verifiable. In the cases of *dṛṣṭa-anumāna* the cognitive passage is from observed to unobserved (yet observable) whereas the one in the case of *sāmānyatodṛṣṭa* is from observed to unobservable. *Dṛṣṭa anumānas,* roughly speaking are analogous to 'scientific inferences.' They operate with our belief in homogeneity of nature and causation. But they do not operate with these beliefs by way of deductive inferences, but they assume the form of analogical or probabilistic inferences. These *anumānas* attempt to provide us with some empirically testable conclusions. *Sāmānyatodṛṣṭa anumānas* on the other hand do not lead to empirically testable conclusions. They are analogous to metaphysical speculations rather than scientific inferences. They are examples of type-crossing inferences because here *pakṣa* and *sapakṣa* belong to different logical types. *Sapakṣa* belongs to the type of those entities which have observable *sādhya*-property. *Pakṣa* belongs to the type of those entities which are claimed to have unobservable *sādhya*-property. Instances in *sāmānyatodṛṣṭa anumānas* function more as 'models' or 'analogies' useful for understanding the *pakṣa*-case than as instances of the same type as the *pakṣa*-case.

§ 8.8

Udyotakara presented his classification of *anumāna* into *anvayi, vyatireki* and *anvayavyatireki* as an interpretation of the ancient classification discussed above (in § 8.7). But in fact it was a development over the old classification. As we saw in § 4.7 this classification is important when we consider *anumāna* as analogical reasoning. *Anvayī hetu* is the signifying analogy which can operate only as a positive analogy and not as negative one in the given *anumāna*. *Vyatireki hetu* is the signifying analogy which can operate only as negative analogy and not as positive one. And *anvayavyatireki hetu* is the signifying analogy which can function both as positive and negative analogy in the given *anumāna*.

But such a classification of *hetu* as signifying analogy does not lead us automatically to an isomorphic classification of *vyāpti*, because the so-called *anvaya-vyāpti* and the so-called *vyatireka-vyāpti* are logically equivalent. (Every *vyāpti* is both *anavyi* and *vyatireki* in this sense.) Such an equivalence does not hold between *anvayi hetu* and *vyatireki hetu* as the two kinds of signifying analogy.

This was suggested first by Dharmakīrti. This was probably his main reason why he was not prepared to accept Udyotakara-style three fold classification of *hetu*. Dharmakīrti, however, introduced a different three-fold classification which suited to the frame of *anumāna* accepted by him.

§ 8.9

According to Dharmakīrti's classification the three kinds of *hetu* are *anupalabdhi* (non-cognition), *svabhāva* (essential nature) and *kārya* (effect). (*NBT* 2.10, 2.11). Non-cognition is used as a *hetu* for inferring a negative fact. When on the other hand a positive fact is to be inferred, there are two alternatives before us. Either *sādhya* and *hetu* constiute the essential nature of one and the same thing such that *hetu* is 'pervaded' by *sādhya* or they are distinct in time but essentially connected with each other because *hetu* is the effect of *sādhya*. These two alternatives correspond to the cases of *svabhāva-hetu* and *kārya-hetu* respectively. They further correspond to two types of *anumāna* viz. *svabhāvānumāna* and *kāryānumāna* respectively. It is possible to interpret the *vyāpti*-statement in *svabhāvānumāna* as expressing the relation of conceptual identity (or analytic necessity) holding between *hetu* and *sādhya*. The *vyāpti*-statement in *kāryānumāna*, on the other hand can be interpreted as expressing 'synthetic necessity' in general and causal necessity in particular. It may be added that Dharmakīrti accepts only one kind of synthetic necessity namely causal. Although it is possible to reinterpret or rather reconstruct Dharmakīrti's classification of *vyāpti* on these lines, Dharmakīrti himself, as I have already suggested in § 5.7 does not put the matter in this way. He does not explain the relation of *tādātmya* which holds between *svabhāva-hetu* and its *sādhya* in terms of partial or total identity of 'concepts' or 'meanings,' but rather in terms of essential nature of things. Dharmakīrti does talk of *sāmarthya* which is supposed to govern the relation between premisses and conclusion of any acceptable *parārthānumāna* or between the statement of *anvaya-vyāpti* and that of *vyatireka-vyāpti*, but *sāmarthya* there stands for logical necessity rather than analyticity.

This opens a different area of study where one could concentrate on the discussions of 'necessity' in Indian logic in general and Dharmakīrti's logic in particular. One can also compare these discussions with their western analogues. This complex job is however not undertaken in this overview.

§ 8.10

We have seen in Chapter V that Dharmkīrti insisted that the *parārthānumāna* should have a valid form. He did this by introducing *vyāpti*-statement as its necessary element. At this stage of Indian logic we come across some forms of inference which closely resemble some of the moods of Aristotalian syllogism. Since Aristotalian logic gives a systematic classification of syllogisms into valid and invalid moods, it could be interesting to see whether different valid and invalid forms of argument discussed by Aristotle can be identified with or reduced to the forms of *parārthānumāna* discussed by Dharmakīrti and the later Indian logicians. Of course the exercise of this kind will deal with many more invalid moods than valid ones. We therefore postpone this exercise to the second part of this chapter. (Please see the sections § 8.27 to § 8.34 and Appendix IV).

In spite of the close resemblance between the forms of *parārthānumāna* in Indian logic and valid moods of syllogism in western logic a fundamental point of difference between Indian and western conception of inference needs to be noted very clearly. Logic for western logicians is an autonomous branch of knowledge which deals with formal validity of arguments or logical truth of the logically true statements as its central principle or norm. It is therefore not concerned with material truth of premises or conclusion as centrally important with regard to acceptability of an argument. Logic, for Indians whether we understand it as *pramāṇa-śāstra* (The science of *pramāṇas)* or *hetu-vidyā* (the knowledge of good or bad reason) or *tarka-śāstra* (the science of reasoning), has been centrally concerned with material truth of a cognition. Material truth of a cognition or true cognition of some extralogical, extralinguistic reality was the main goal of all *pramāṇas* including *anumāna*. The acceptability of any inference therefore necessarily implied the material truth of its premises and conclusion. Even when *vyāpti* was introduced as a necessary element of inference and hence formal validity was made a necessary condition of the acceptability of any *parārthānumāna*, formal validity was important because it served as a means to true conclusion (- of course even as that it was only a necessary condition and not the sufficient one -), it was never important as an end in itself. This approach of Indian logicians must have helped them to put inference always in a broader framework of our search after knowledge of the Reality. But at the same time it perhaps had a disadvantage namely that formal logic did not grow in India as an independent inquiry into logical truth.

Part II : Fallacies
§ 8.11

While considering the fallacies discussed in Indian logic, the term *hetvābhāsa* is generally regarded as synonymous of the term fallacy. Again the term *hetvābhāsa* is generally taken in the sense attached to it by Nyāya logicians. But Nyāya notion of *hetvābhāsa* is so vague (or one may say, exhaustive) that in Nyāya frame we cannot distinguish between *pakṣābhāsa*, *hetvābhāsa* and *dṛṣṭāntābhāsa*. Nyāya logicians thought that the fault in any element of inference was directly or indirectly related to the *hetu*-element and therefore all the fallacies can be reduced to the faults in *hetu*-element.

This attitude of Nyāya logicians provided a crude way of bringing all varieties of fallacies under one title. But the attitude was not only orthodox but also misleading. The founder of Nyāya-school, Akṣapāda, did not mention either *pakṣābhāsa* or *dṛṣṭāntābhāsa*. Therefore the Nyāya logicians after Akṣapāda, who were made aware by Diṅnāga of *pakṣābhāsas* and *dṛṣṭāntābhāsas*, generally neglected them or tried to reduce them to *hetvābhāsa*[11].

The attitude was misleading because a thesis or an instance presented in an argument can be faulty irrespective of the *hetu* provided for the proof of the thesis being faulty. Therefore Buddhists seem to be right in distinguishing between fallacies of *pakṣa, hetu* and *dṛṣṭānta*.

The term *sādhana* is ambiguously used in Indian logic and is many times confused with *hetu*. Naiyāyikas especially do not distinguish between *sādhana* and *hetu*. But Buddhist logicians are generally found particular about the distinct use of the term *sādhana*. Especially Diṅnāga clearly distinguishes between *sādhana* and *hetu*. For him, *sādhanc* stands for argument and *pakṣa, hetu* and *dṛṣṭānta* stand for the elements of argument. Diṅnāga was aware that the fault in any one element of argument does result into a fault in argument, i.e. in *sādhanābhāsa*, but does not necessarily result into the fault in other elements of the argument. Therefore Buddhist logicians regarded the fallacies of *hetu, pakṣa* and *dṛṣṭānta* as the types of *sādhanābhāsa* and did not identify *hetvābhāsa* with *sādhanābhāsa*. Even Dharmakīrti is not particular about identifying the two.

§ 8.12

Sādhanābhāsa means an argument which appears to be sound but is not sound.[12] The term fallacy is similarly defined as argument which seems to be valid but is not so.[13] However, if we want to include material fallacies also in the list of fallacies, then we have to define fallacy as a seemingly sound but actually unsound argument. Thus the term *sādhanābhāsa* appears to be synonymous with 'fallacy.'

There is another reason why *sādhanābhāsa* would be a better synonym for fallacy than *hetvābhāsa*. In the treatises of Indian logic (starting from that of Diṅnāga) *anumāna* is classified in two types. *Anumāna* which occurs to oneself and *anumāna* which is made to justify one's own position to others. When *anumāna* is made for justifying one's position to others, it takes the form of argument. When on the other hand *anumāna* occurs to oneself, it may take the form of a non- linguistic cognition. Now, the concept of fallacy can apply only to an argument and not to a non-linguistic cognition. The term *hetvābhāsa*, on the contrary, applies to both the types of *anumāna*. The term *sādhanābhāsa* as used by the Buddhist logicians, on the other hand, refers only to a fault in argument and does not refer to a fault in *anumāna* which occurs to oneself.

§ 8.13

While discussing the concept of fallacy in ancient Indian logic, the term *ābhāsa* is very important. *Ābhāsa* of x does not merely mean non-x. Thus fallacy does not mean merely invalid argument or unsound argument, but the fallacious argument should nevertheless appear like a sound argument. This thread of thought is sometimes lost in the discussions of fallacies and the classification of fallacies is made solely on the basis of the forms of unsoundness of an argument. In Indian logic the terms *anapadeśa* and *alinga* are many times used to connote fallacies. For instance the term *anapadeśa* occurs in *VS* 3.1.10. *Anapadeśa* means non-*hetu*. Thus any particular *anapadeśa* need not necessarily resemble closely or appear as a sound *apadeśa*. A good example of this is the strict (non-contextual) *viruddha*. In case of strict *viruddha* the *hetu* in fact implies the negation of *sādhya*, though it is argued to imply the *sādhya*. In *VS* this fallacy is called as *asan*. Kaṅāda gives the following example of this fallacy : 'This must be horse because it has horns.' Here, 'having horns' does not appear to be sound *hetu* even to a man of minium general knowledge. In this sense *asan* may be called as *anapadeśa*, but not as *hetvābhāsa*. This *asan* or *viruddha* of Kaṅāda can be compared with the *viruddha hetvābhāsa* of *NS*. *Viruddha* of *NS* which is a contextual fallacy appears as a good *hetu* on the surface of the argument. Thus *viruddha* of *NS* is not only *anapadeśa* but also *hetvābhāsa*. This distinction between *hetvābhāsa* and *anapadeśa* which was clear to the earlier logicians of Nyāya as well as Vaiśeṣika school, was neglected by the later logicians starting from Diṅnāga. It was on the contrary held by later logicians that any unsound argument resembles and therefore may appear as a sound argument, because it is equally an argument. As far as we can find out some similarity between a sound and an unsound argument we need not distinguish between *hetvābhāsa* and *anapadeśa*.

If we call appearance-aspect of fallacies as the positive aspect and the unsoundness-aspect of fallacies as the negative aspect, then we may say that Akṣapāda and Vātsyāyana emphasised both the aspects, while first Kaṇāda and then Diṅnāga emphasised the negative aspect. All the Indian logicians after Diṅnāga followed this trend of emphasising the negative aspect.

§ 8.14

Though *sādhanābhāsa* is a better synonym for fallacy, Buddhists' treatment of *sādhanābhāsas* appears to be different from the western treatment of fallacies. While discussing *sādhanābhāsa*, Buddhists are more interested in the material fallacies rather than in the formal fallacies. But for western logicians, formal fallacies are more basic than material fallacies. Many times western logicians define fallacies as invalid arguments which means formal fallacies. This is in tune with their differential approaches to inference (See § 8.10)

However, the charge that is put forth against Indian logicians that Indian logicians never considered any formal fallacies at all, is wrong. Though separate and special attention was not paid to the notion of formal fallacies, still some Indian logicians did recognise some forms of formal fallacies. Dharmakīrti was mainly responsible to take the note of some formal fallacies.

§ 8.15

There are some notions which are not explicitly brought under the head *sādhanābhāsa* or *hetvābhāsa*, but which in fact may be treated as the species of fallacies. *Chala jāti* and *nigrahasthāna* are some such major notions. In some books on Indian logic all these terms are said to connote the types of fallacies.[14] This view, however,requires examination. It seems to me that *chala* and *jāti* can be included in fallcies, but all the *nigrahasthānas* cannot be so included.

§ 8.16

For the examination of *chala* as *sādhanābhāsa*. Let us take the following dialogue.

Mr. A : I had ordered you to save soap and waste-paper.

Mr. B : Yes sir. I am obeying your orders because I am saving soap and wasting paper.

Mr. A : I had not told you to waste paper but to save waste-paper.[15]

In the above dialogue Mr. B is making a *chala*- argument. The conclusion of his argument is that he is acting according to the orders of Mr. A. And his premises can be stated as follows :

1) For all x if x saves soap and wastes paper, then x acts according to the orders of Mr. A. (supposed premise)

2) I (Mr.B) am saving soap and wasting paper.

Mr. A while opposing Mr. B 's argument points out that the *vyāpti*-statement assumed by B, which is the suppressed premise of B's argument, is false.

From the above example it may become clear that *chala* type of argument can be identified with *hetvābhāsa*.[16] *Hetvābhāsa* being included in *sādhanābhāsa*, *chala* can be included in *sādhanābhāsa*.

§ 8.17

Similar to *chala, jāti* also can be included in *sādhanābhāsa.* Generally *jāti* is a counter-argument attempted by the opponent to prove a certain result which is unacceptable to the disputant. Many times the opponent's *jāti*-argument is the counter-argument based on mere positive or negative analogy. Thus either such argument,being analogical, does not include the necessary *vyāpti*-statement, or the *vyāpti*-statement which is included by the opponent as a premise of his counter-argument is false. In the first case *jāti*-argument is a formal fallacy (which may be included in *ananvaya* or *avyatireka*, or *nyūna*, all of which are the types of *sādhanābhāsa*). In the second case *jāti*-argument is a case of *hetvābhāsa.* Vātsyāyana while answering the first two *jātis* claims that the *jāti*-arguments cannot stand because they are the cases of *hetvābhāsa.*[17]

Buddhist logicians, however, have given the distinct accounts of *sādhanābhāsa* and *dūṣaṇābhāsa* (= *jāti).* Thus Buddhists seem to distinguish between fallacies of proof and fallacies of disproof. But it may be seen that disproof also is a kind of proof. Proving p may mean proving p or proving necessarily p, while 'disproving (refuting) one's thesis p' may mean proving not-p or proving necessarily not-p or proving 'possibly not-p'. Thus, if the disproof of p is fallacious, then this means that the proof of not-p or that of necessarily not-p or that of possibly not-p is fallacious. There seems therefore to be not harm in including *dūṣanābhāsa* (= *jāti*) in *sādhanābhāsa.*

§ 8.18

Though sometimes the scholars of Indian logic are inclined to include *nigrahasthāna* in the list of fallacies, it should be noted, as we have seen (in § 7.62 etc.) that all the types of *nigrahasthāna* are not the types of fallacies. Even though we distinguish between categorically proper, conditionally proper and improper occasions of defeat, even all the categorically proper occasions of defeat are not fallacies. However some of the categorically proper occasions of defeat can be regarded as fallacies. (See § 7.62).

§ 8.19

The other species of fallacies which are not traditionally included in *sādhanābhāsa* are *prasaṅga-* fallacies, *nayābhāsas* and the fallacies of definition.

In all the cases of *prasaṅga*-fallacies it is clear that while discovering these fallacies in the arguments, arguments are not directly shown to be unsound; it is only shown that if these are accepted as sound, some unacceptable consequences may follow. Thus the argument is indirectly shown as unsound. It can be held therefore that the *prasaṅga*-fallacies do not differ from *sādhanābhāsas* as far as the unsoundness of these fallacious arguments is concerned. The difference between the two lies in the method of discovering the fallacies.

§ 8.20

Though *nayavāda* and *nayābhāsa* both refer to descriptions and not to arguments, still the descriptions in the case of *nayābhāsa* should be considered as the conclusions of the enthymematic arguments. If for instance a philosopher states his thesis that 'substance and quality are utterly distinct from each other' and thus commits *naigamābhāsa*, then while stating the thesis the argument that he intends to make is as follows :

'If any two words are not synonymous, then they denote utterly distinct entities. For instance 'table' and 'chair.' 'Substance' and 'quality' are also non-synonymous words. Therefore the words denote utterly distinct entities.'

In the above argument the first premise is false and therefore the argument is unsound. In a similar way all the cases of *nayābhāsa* can be shown as the cases of 'fallacy' i.e. *sādhanābhāsa*.

§ 8.21

When somebody asserts that D' is a definition of D, he assumes that the following premises imply his assertion,

1) The relation between 'x has D' and 'x has D''is that of material equivalence for all values of x.

2) The clearcut knowledge of D' is practically useful for knowing D clearly and distinctly.

3) Whenever such two conditions (which are suggested by premises 1) and 2) are satisfied by the definiens, the definiens is supposed to give a good definition of the definiendum.

The third premise gives us the general definition of definition, while 1) and 2) give us the information as to how the conditions of definition are fulfilled by D and D.' The conclusion of such argument would be then that D' gives a good definition of D.

Now the fallacies *avyāpti, ativyāpti* and *asambhava* are the falsifications of the first premise. While *itaretārāśraya* is the falsification of the second premise. Thus there seems to be no harm in regarding the 'fallacies of definition' as *sādhanābhāsas*.

§ 8.22

The fallacies can be classified in many ways. They are, for example, classified in western logic under formal fallacies and material fallacies. This classification was introduced by Whately as against the Aristotalian classification into fallacies dependent on language and fallacies outside language. Whately brings the first type recognised by Aristotle under 'formal fallacies' and calls them semilogical fallacies.[18] Though we do not have synonymous classificatory terms used in ancient Indian logic, we notice that Bhāsarvajña made a distinction between syntactical fallacies and semantical fallacies.[19]

It should be noted here that the distinction between formal and material fallacies is not identical with the distinction between syntactical and semantical fallacies. For instance, starting from Whately, the so called semilogical fallacies, which are in fact the fallacies of inattention to the ambiguities of language, are included by Whately in formal fallacies. It may be noted, however, that these fallacies dependent on language can be regarded as formal fallacies if interpreted in a certain way and as material fallacies if interpreted in certain other way.

§ 8.23

Let us consider an example of 'accident.'

'What you brought yesterday, you eat today.

You brought raw meat yesterday.

Therefore you eat raw meat today.'

In Indian logic such a case is called *sāmānyacchala*. Here the first premise is ambiguous. It may mean (1) Generally what you brought yesterday, you eat today (after processing it whenever necessary). Or (2) Without any process of cooking etc., every time what you brought yesterday, you eat today.

If the first interpretation of the first premise is accepted, then the conclusion does nto logically follow from the premises. Thus the argument in this case should be regarded as a formal fallacy. If the second interpretation of the first premise is accepted, then though the argument then is formally valid, the first premise is false. Thus the argument is materially fallacious.

But formal fallacy is not the same as syntactical fallacy and material fallacy is not the same as semantical fallacy. And when we consider the classification of fallacies in terms of syntactical and semantical fallacies, the same example of accident would be brought under semantical fallacy. We can see that the example of *sāmānyacchala* clearly appears to be valid, when we consider it syntactically, i.e. when we consider only the surface structure of the propositions. Thus it would be proper here to claim the argument to be semantically fallacious.

§ 8.24

An example of secundum quid is more problematic, which runs,

1) This dog is yours.

2) This dog is father.

Therefore this dog is your father.

It may seem that the argument is syntactically good. But actually it is not so. If the conclusion was 'This dog is yours and father,' the argument would have been syntactically good. As it is given, the argument also involves quibbling. Thus it raises the question whether quibbling should be regarded as syntactical fallacy or a semantical fallacy. It may be claimed here that some examples of quibbling are syntactical fallacies and the others are semantical fallacies.

An explanation of the example may lead us to regard the example as a case of semantical fallacy. Here we interpret the first premise as 'Whatever the dog is, is that in relation to you' (Thus if the dog is a servant, it is your servant, if the property, your property and so on, in case of any relational predicate) And then the further steps follow,

...This dog is father

The dog is father in relation to you.

(i.e., The dog is your father.)

In fact here the interpretation of the first premise as 'Whatever the dog is, is that in relation to you' amounts to be a kind of *nayābhāsa*. Because one who intends to make such an interpretation, neglects the contextual meaning of the sentence and tries to make a context-free application of the sentence. The first premise namely 'This dog is yours' is a *nayavāda*, an ordinary language sentence, which has a contextual application. To interpret it as 'In a certain respect the dog is yours, but in certain other respect the dog is not yours' would be to make use of *syādvāda*. But to interpret the premise as 'In all respects the dog is yours' is to commit *nayābhāsa*. (In this way fallacies like 'accident' and 'secundum quid' can be viewed as the cases of *chala* or *nayābhāsa*.) However, it may be suggested here that the argument as the example of secundum quid is a formal fallacy when we consider the *syādvāda*-interpretation of the first premise, and it is a material fallacy if we consider the *nayābhāsa*-interpretation of the first premise. In both the cases the argument involves syntactical fallacy. It involves semantical fallacy only when we consider *nayābhāsa*-interpretation. Thus under the *nayābhāsa*-interpretation, the same example is an example of both, syntactical fallacy and semantical fallacy.

Here, for getting rid of the problem, the principle that should be adopted is that though the same case may be the case of many fallacies, the same fallacy cannot be many fallacies. Now what is *nayābhāsa* (or *chala*) ? Is it a syntactical fallacy or a semantical fallacy ? It cannot be both.

§ 8.25

We can use a decision-procedure to decide as to whether a particular fallacy is syntactical or semantical, which may be stated as follows.

Given a fallacy we take any good example of the fallacy. If the example is not syntactically a valid argument, then we try to convert the syntactically invalid argument into the syntactically valid argument and at the same time retain the fallacy of which it is the example. Now we take the decisions in the following manner.

1) If the fallacious argument is already syntactically valid, then the fallacy of which it is an example, is a semantical fallacy.

2) If the fallacious argument is not syntactically valid but can be converted into a syntactically valid argument without replacing the particular fallacy in it, then also the fallacy of which it is an example, is a semantical fallacy.

3) If we cannot change the argument into a syntactically valid argument and yet retain the fallacy, then the fallacy must be syntactical.

Again we come to the problematic example of secundum quid. We can convert the example into following.

1) The dog is yours.

2) In other words, whatever the dog is, is that in relation to you.

3) The dog is a father.

Therefore the dog is your father.

Now the argument is syntactically valid. But the second premise ('In other words....') is obviously false due to *nayābhāsa*, or secundum quid. Thus the fallacy is retained in spite of the conversion of the argument into a syntactically valid one. In a similar way all the semilogical fallacies can be shown to be semantical fallacies.

§ 8.26

By using the procedure suggested in § 8.25 we can find out that all the fallacies which come under *hetvābhāsa*, *chala*, *nayābhāsa*, *nigrahasthāna*, fallacies of definition and *prasaṅga* are semantical fallacies while *nyūna* (as accepted by Dharmakīrti), *ananvaya*, *avayatireka*,[19a] *viparītānvaya* and *viparītavyatireka* are all syntactical fallacies.

Though the *jāti*-types of arguments as instanciated by Nyāya logicians, are presented loosely, so that one may feel them to be enthymematic or formally invalid, still what is generally held by the Nyāya logicians is that syntactically *jāti*-argument can be of the same nature as a sound argument. A *jāti*-argument can be a formally valid argument and still may be based on mere analogy, because the debater who is using a *jāti*-argument may make a necessary *vyāpti*-statement, which he himself knows to be false. His opponent can then well argue that the debater's argument is based on mere analogy (though syntactically it is not 'an analogical argument') and the genuine true *vyāpti*-statement necessary for his argument cannot be formulated. Thus the *jāti*-arguments can be converted into syntactically valid arguments and still their fallacious nature can be retained. Therefore *jātis* should be regarded as semantical fallacies.

§ 8.27

BODAS in *TS* suggests that there are no purely formal fallacies discussed in Indian logic.[20] It seems that BODAS was mainly engaged in Navya-Nyāya logic and was unaware of the formal fallacies discussed in ancient Indian logic. In spite of his view that in Indian logic there are no formal fallacies discussed, BODAS undertakes a programme[21] of converting various western examples of invalid fallacious arguments in to Sanskrit *anumānas* and pointing out the Nyāya fallacies in those examples. In his programme he generally tries to demonstrate the formally fallacious arguments discussed in western logic to be the examples of materially fallacious arguments of Indian logic. His programme becomes misleading because though it is possible that some of the formally fallacious argument are also materially fallacious, we should be able to point out the Indian formal fallacies in the examples of western formal fallacy.

§ 8.28

The important difference between the moods we have in Indian logic and those in western logic, seems to be that in Indian logic all the moods, valid or invalid, end with the conclusion which is A-proposition. The western syllogisms can end in any type of conclusion, A, E, I or O. The E- propositions can be converted into A-propositions, but I or O-propositions cannot. Thus the arguments which conclude in I-proposition or O-proposition, cannot be shown to be valid or invalid in Indian way.[21a] But the arguments which conclude in A-proposition or E-proposition can be examined and demonstrated to be valid or invalid, whatever they are, by comparing them with the model valid or invalid arguments discussed in Indian logic. In the following sections § § 8.29-31 we shall consider a procedure by which we shall be in a position to point out an Indian formal fallacy in the argument which is formally fallacious according to western logic. Here of course we have to leave the arguments which conclude in I or O- proposition, unexamined. (By pointing out the 'Indian formal fallacy' I only mean pointing out the Indian name for the formal fallacy).

§ 8.29

We take any concerned syllogism in any mood or figure with S as the subject of the conclusion, P as the predicate of the conclusion and M as the middle term.

First Step :

We start with examination of the conclusion. We try to translate the conclusion into A-propostion of the form,

All S is X.

In trying to translate, we come across either of the three possible cases -

(α_1) : We cannot translate so. This would occur when the conclusion is either I-proposition or O-proposition.

(α_2) : X = non-P. This would occur if the conclusion is E-proposition.

(α_3) : X = P. This would occur if the conclusion is already A-proposition.

Decisions : In the case of α_1 we decide that we cannot examine the argument to be either valid or invalid in the frame of Indian logic.[21a]

In the case of α_2 and in case of α_3 we decide that X is the *sādhya*-predicate of the argument and then turn to the second premise.

§ 8.30

Second Step :

We try to translate the second premise into A-proposition of the form

All S is Y.

Here we come across any of the three possible cases.

(β_1) : We cannot translate so. This would occur if the given second premise is either I-proposition or O- proposition or the proposition of the form

All M is S.

(β_2) : Y = non-M. This would occur when the given second premise is E-propositon.

(β_3) : Y = M. This would occur when the given second premise is of the form

All S is M.

Decisions : In the case of β_1 the argument suffers from illicit process of minor because S is distributed in the conclusion but not in the premises. From the point of view of Indian logic the argument suffers from '*nyūna* due to the lack of *pakṣadharmatā*-statement' because there the *hetu* is not shown to be predicated of the pakṣa-subject namely 'All S.'

In the case of β_2 and in the case of β_3 we decide that Y is the *hetu* of the argument and then we turn to the first premise.

§ 8.31

Third Step :

We try to translate the first premise in A-proposition. While translating it we are not allowed to transpose it, so that we can point out the exact fallacy, if any. Here, while translating we come across either of the following five possible cases.

(γ_1) : We cannot translate the first premise into A-proposition. This would occur in case the premise is I-proposition or O-proposition.

(γ_2) : We can translate the first premise in any of the two forms.

All Y is X

All non-X is non-Y

(γ_3) : We can translate the first premise in the form,

All X is Y

(γ_4) : We can translate it in the form,

All non-Y is non-X.

(γ_5): We can translate it in any of the following forms,

All X is non-Y

All non-X is Y

All non-Y is X

All Y is non-X.

Decisions : In case of γ_1 as well as in case of γ_5 we find that we cannot translate the first premise into a proper *vyāpti*-statement. Thus we conclude that the argument suffers from '*nyūna* due to the lack of *vyāpti*- statement.'

In the case of γ_2 we decide that the given argument is valid.

In the case of γ_3 we find that the outcome of the translation is the converse of the proposition of *anvayavyāpti* because the *anvayavyāpti* should be stated as

All Y is X.

Therefore in the case of γ_3 we decide that the argument suffers from the fallacy called *viparītānvaya*.

In the case of γ_4 we find that the outcome of the translation is the converse of the proposition of *vyatirekavyāpti* because the *vyatirekavyāpti* should be stated as

All non-X is non-Y.

Therefore in the case of γ_4 we decide that the argument suffers from the fallacy called *viparītavyatireka*.

§ 8.32

Though in the above-stated procedure, for the sake of convenience we have regarded *nyūna* as a distinct fallacy from other formal fallacies, *nyūna* is in fact the general term to connote any invalid argument, because *nyūna* means insufficient reasoning. Thus, as the proper *vyāpti*-satement which is necessary for making the argument valid, is lacking in the case of *viparītānvaya* or *viparītavyatireka*, we can call these as the subtypes of *nyūna*. Thus we conceive of at least four types of *nyūna*.

1) *Nyūna* due to the lack of *pakṣadharmatā*- statement. (The case of β_1, see § 8.30).

2) *Nyūna* due to the lack of *vyāpti*-statement or its converse (The case of γ_1 or γ_5, see § 8.31).

3) *Viparītānvaya* (The case of γ_3, see § 8.31)

4) *Viparītavyatireka* (The case of γ_4, see §8.31)

§ 8.33

Following the procedure stated in §§ 8.29-31 we can show that any case of illicit process of minor is the case of '*nyūna* due to the lack of *pakṣadharmatā*- statement,' but not vice versa. Supposing that we have an argument of the form

All M is P

Therefore All S is P,

then from the point of view of Indian logic, the argument lacks the *pakṣadharmatā*-statement and therefore suffers from the fallacy of '*nyūna* due to the lack of *pakṣadharmatā*-statement.' But in western logic this would not be called the case of illicit process of minor, but the argument will be supposed to be no syllogism (of course presuming that it is not an enthymeme). So we may say that the argument suffers from the fallacy of two propositions.

§ 8.34

For the sake of exemplification we can take two examples of invalid arguments considered by BODAS and apply our procedure.

I 1) Some things that glitter are not gold.

2) Glass is a thing that glitters.

Therefore Glass is not gold.[22]

This argument is a case of undistributed middle. BODAS translates the first premise as

यत्र यत्र तेजस्वित्वं तत्र तत्र सुवर्णत्वाभावः ।

This is obviously a wrong translation. In fact the translation should be as

1) यत्र यत्र तेजस्वित्वं तत्र तत्र न सुवर्णत्वम् ।

or 2) क्वचित् तेजस्वित्वं न तु सुवर्णत्वम् ।

However, when we apply our procedure for the examination of the argument, we find that *sādhya*-predicate of the argument which we symbolized as X is 'the property of not being gold.'

The *hetu* of the argument which we symbolized as Y is 'the property of glittering.' At the third step of the procedure we find that we cannot translate the first premise into A-statement because it is O-statement. Thus we come across the case of γ_5 and conclude that the argument suffers from *'nyūna* due to the lack of *vyāpti*-statement.'

II 1) All books are liable to err.

2) All books are human productions

All human productions are liable to err.[23]

Here BODAS translates the first premise as 'पुस्तकत्वात्' This is obviously a wrong translation. The correct translation would be यत्र यत्र पुस्तकत्वं तत्र तत्र प्रमादार्हत्वम् ।

When we apply our procedure for the examination of the argument, we find that the *sādhya*-predicate which we symbolize as X is 'the property of being liable to err.' At the second step we find that this is the case of β_1 because the second premise is of the form

All M is S

Therefore we conclude that the argument is fallacious due to *'nyūna* due to the lack of *pakṣadharmatā*-statement.'

In the above examples (I and II) and the others cited by BODAS in *TS* pp. 321-6, BODAS tries to point out the Indian types of material fallacies in the cases of formal fallacies. But there can be some arguments where we find only formal fallacies but not material fallacies. In such cases BODAS's programme would fail. Dharmakīrti seems to be aware of such cases when he distinguishes between *ananvaya* and *apradarśitānvaya* and also between *avayatireka* and *apradarśitavyatireka*. (See § 5.28)

8.35

Two other strictly formal fallacies that are discussed in western logic are named as 'Affirming the consequent' and 'Denying the anticident.' 'Affirming the consequent' and 'Denying the anticident' apparently resemble *viparītānvaya* and *viparītavyatireka* respectively.

Take for instance, a case of 'Denying the anticident'

"1) If Bacon wrote Hamlet then Bacon was a great writer

2) Bacon did not write Hamlet.

Therefore Bacon was not a great writer."

This appears to be a case of *viparītavyatireka* because the argument might be translated in the following way so that it will appear a case of *viparītavyatireka*.

(I) If anybody wrote Hamlet, then he was a great writer.

Bacon did not write Hamlet.

Therefore he was not a great writer.

Or (II) If anybody was Bacon and wrote Hamlet, then he was a great writer.

Mr. X was Bacon, but he did not write Hamlet.

Therefore Mr. X was not a great writer.

It should be noted, however, that such an attempt to translate a statement about Bacon in a statement about anybody in the world, is artificial. Because the statement namely 'If Bacon wrote Hamlet then Bacon was a great writer' is not a *vyāpti*-statement although it is an instançiation of a universal conditional.

§ 8.36

As we already saw that the so-called semilogical fallacies can be included in semantical fallacies (See § 8.15); the well-known fallacies dependent on language namely Amphiboly, Equivocation, Accent, Figure of Speech and Composition and Division are all semantical fallacies. We do not find the exact Indian parallels to all of them, though it does not mean that committing some of them was sanctioned as an innocence by Indian logicians. Amphiboly and Equivocation can be regarded as the subtypes of *vākchala*. Any exact parallel is not found in Indian logic for Accent. But in a sense, when '*vāk*' means utterred sentences, Accent may be regarded as a subvariety of *vākchala*[23].

When 'Figure of Speech' means a fallacy which occurs due to the confusion between figurative meaning and literal meaning, it closely resembles *upacāracchala*. But Figure of Speech is also interpreted as the confusion between etymology or grammatical forms and the nature of things.[24] We see that in cases of *śabdanayābhāsa*, *samabhirūḍhābhāsa* and *evambhūtābhāsa* such a confusion is present. *Śabdanayābhāsa* occurs due to the confusion between grammatical forms of the words and referencial meanings of them. The other two *nayābhāsas* are committed due to the confusion between the etymological meanings and the conventional referencial meanings of the words. (See §§ 7.89-91)

The fallacy of Composition and the fallacy of Division are nowhere noticed in Indian logic. However, some other philosophical ideas show the insight of Indian philosophers, in the nature of these fallacies. In Advaita-Vedānta and Nyāya, for instance, the distinction between *aṁśa* and *avayava* is held important.[25] *Aṁśa* refers to that part of a whole *(aṁśi)* which has the same properties which the whole has.

For instance, a drop of water also is water. But *avayava* (which literally means limb) is that part of the whole *(avayavī)* which does not have the same properties as the whole has. For instance, a hand of a man is only *avayava* of a man, a hand of a man is not a man. The distinction clearly shows that in some cases a member of the class does not bear the properties the class bears and vice-versa. In some cases, however, the properties of a class and the properties of a member of the class are identical.

Of course the idea of non-distinction between *aṁśa* and *aṁśī* should not be stretched too far, because *aṁśa* and *aṁśī* also are, in a way, distinguishable from each other. Taking the same example we can say that the class of hundred 'drops' of water is not again a 'drop' of water.

§ 8.37

There are some other fallacies which occur due to inattention to ambiguities of language, like the so-called Boethian fallacies. Our ordinary language has some natural deficiencies and as HAMBLIN points out, the fallacy of Secundum Quid is an everpresent and unavoidable possibility in practical situations.[26] The so-called Boethian fallacies of Different Part, Different Relatum, Different Time and Different Modality[27] are also of the same kind as Secundum quid. These fallacies can be well brought under *nayābhāsa*. Jaina logicians while analysing the descriptions with the help of *syādvāda* point out that a description of a thing is true only in relation to that particular thing *(svadravya)*, the particular place (*svakṣetra*), the particular time *(svakāla)* the particular modality *(svabhāva)*. But in ordinary language, when we describe or come across different descriptions we do not mention or hear others mentioning the exact contexts (the thing-context, the place-context, the time-context and the modality-context) in which the description is true. Practically also it is difficult. The description thus leaves the lacuna for others to imagine the contexts in which the description gets falsified. In this way our ordinary language is ambiguous and one who does not attend the proper contextual meanings of the descriptions, may commit the Boethien fallacies. The Boethian fallacies thus are the clear breaches of the principle of *syādvāda*.

For instance if somebody claims that the conjunctive statement 'This eye is white and this eye is also non-white' is contradictory, commits the Boethian fallacy of Different Part. A Jaina logician would say that his claim amounts to a kind of *nayābhāsa* and the seeming contradiction in the conjunctive statement should be removed by applying the term *syāt* (= in a certain respect) to each conjunct.[28]

§ 8.38

We can thus account for the semilogical fallacies, i.e. the fallacies of inattention to the ambiguity of language. As we saw, these account for semantical fallacies. The other types of semantical fallacies are the material fallacies of Whately's classification. In this group of fallacies we may include

1) Fallacies of falsity or dubitability of some step in the argument.

2) Fallacies of irrelevant conclusion (Ignoratio Elenchi).

§ 8.39

In the context of Indian logic, first type stated in § 8.38 may be again classified into three subtypes.

A) Falsity or dibitability of the *pakṣadharmatā*- premise.

B) Falsity or dubitability of the *vyāpti*-premise.

C) Falsity of dubitability of the conclusion (shown by some other means when A or B cannot be directly shown).

We can easily see that various subtypes of *asiddha* (excluding *anyathāsiddha*) would come under (A); various subtypes of *anaikāntika* (excluding *viruddhāvyabhicāri*), *viruddha* and also *anyathāsiddha* would come under (B); various subtypes of *bādhita* and also *satpratipakṣa* (or '*viruddhāvyabhicāri*') would come under (C).

We do not find an identical classification in western logic, because the inferences in western logic are not necessarily discussed by using the notions identical with *pakṣadharmatā* and *vyāpti*. Thus we can have the common head for (A) and (B) called 'the fallacies of premise unduely assumed'.[29] Out of the two, the fallacies based on 'falsity or dubitability of the *vyāpti*-premise' have another aspect by which they are regarded as inductive fallacies. We shall discuss these in §§ 47-56

Prasaṅga-fallacies also are the fallacies based on falsity of some step.

§ 8.40

Petitio principii is a material fallacy where a premise or premises depend on the conclusion, for their own proof. That is to say, in the case of petitio, premises can be known to be true only if the conclusion can be known to be true. But the case of any argument is that the conclusion is known to be true only when the premises are known to be true. Thus the premises in such a fallacious argument remain in want of proof. In this way the premises in such an argument are unsupported or uncertain. Petitio principii, therefore, may be regarded as a fallacy of the dubitability of premises.

In the context of Indian logic, *prakaraṇasama* of Akṣpāda comes nearest to the description of petitio-principii.[30] It may appear, however, that the *prasaṅga*-fallacies called *ātmāśraya, itaretarāśraya* and *cakraka* are the varieties of petitio principii. For instance when the thesis Y is proved on the basis of X and X is to be proved on the basis of Y, then the argument namely X /∴Y, is fallacious due to *itaretarāraya*. But it should be noted here that though we call this fallacy *itaretarāśraya*, it would not be a *prasaṅga*- fallacy, because in the case of the *prasaṅga*-fallacy, of circular dependence, we should be 'led' to some circular position if we accept the argument. But in the case of petitio we are not 'led' to circularity by the argument, but the argument itself is noticed to be circular. Thus arguing in circle (= petitio principii) should be distinguished from the *prasaṅga*-fallacy of circular dependence/self-dependence. In ancient Indian discussions in philosophy, sciences etc., the terms *ātmāśraya*, *itaretarāśraya* and *cakraka* are used to denote *prasaṅga*-fallacies and not to denote petitio-principii.

Though petitio principii, in a way, can be reduced to epistemic types of *ātmāśraya* /*itaretarāśraya*/*cakraka*, we would not be concerned there with the epistemology of cognition but with the epistemology of justification. In epistemology of justification we are not concerned with the causal relation between one cognition and the other, but we are concerned with justifiability of one cognition in terms of the other. In epistemology of cognition, however, we are concerned with the causal relation between one cognition and the other. That is why, as I pointed out in the case of mutual dependence,[31] the epistemic types of *ātmāśraya*, *itaretarāśraya* and *cakraka* can be reduced to the corresponding 'existential' types. But these fallacies considered as the varieties of petitio-principii, cannot be reduced to existential types. This distinction is to be kept in mind when we try to equate *ātmāśraya*, *itaretarāśraya* etc., with the varieties of petitio principii.

Another distinction that should be noted is that, the premises of a circular argument are unsupported (because they are 'supported' only by their conclusion), but may not be false. But a *prasaṅga*-fallacy is an argument, a premise or the conclusion of which is shown to be false, by demonstrating that it leads us to some unacceptable (false) consequence.

§ 8.41

As compared with the *prasaṅga*-fallacy called *anavasthā*, in western philosophy we have the fallacy of infinite regress (also called ad infinitum). For instance in accepting God as unmoved mover it is claimed that if every mover is moved by some other mover then we will be led to infinite regress. Therefore we have to accept some mover who is not moved by any other mover.

However, the fallacy of infinite regress or ad infinitum does not find mention in any western official list of fallacies. Though in fact the analogues of *prasaṅga*-fallacies are not mentioned in any official texts of western logic, they are relevant in the context of the method of reductio ad absurdum which is found discussed in western logic.

§ 8.42

Under the second type of fallacies in western logic (namely 'fallacies of irrelevant conclusion') stated in §8.38, we have the varieties of Ignoratio Elenchi such as Ad hominem, Ad verecundiam, Ad misericordiam, Ad ignorantiam etc. We do not find exact parallel for any of these in Indian logic. However, indirect awareness about some of them can be found.

Ignoratio elenchi means ignoring the issue or misconception of refutation.[32] This is something like the occasion of defeat called *arthāntara*. But there is a difference. In the case of ignoratio elenchi, the arguer makes the irrelevant points for proving his conclusion. In the case of *arthāntara*, the arguer makes irrelevant points but they may or may not be supposed to prove the respective conclusion; the arguer may be only intending to make a diversion. Whately brings the subtype called 'fallacy of shifting ground' under Ignoratio elenchi.[33] This subtype can be noticed in some cases of *hetvantara*. It may also be noticed that the tendency of ignoring the issue, which is the essence of ignoratio elenchi, is clearly seen in various types of *jāti*.

Ad hominem is committed when a statement is accepted to be true on the basis of the motives or background of its supporter or opponent.[34] Ad verecundiam is committed when a statement is accepted to be true on the basis of the authority of its supporter.[35] We find no mention of any of these fallacies in any Indian treatises. But one may argue that Indian philosophers are used to commit these fallacies, especially the second. It may be pointed out that while defining *āpta* (= an authentic person), not only his unambiguous knowledge and attitude of speaking truth is asserted but also his moral attitude is affirmed.[36] It may be suggested therefore that claiming the authenticity of a statement on the basis of the moral conduct of its supporter, is in a way, commi.ting ad hominem.

However, such an account of ad hominem is dubitable because, as said by Joseph, man's character bears upon his credibility.[37]

The excessive respect for authority which causes ad verecundiam is, however, observed to be a salient feature of traditional Indian philosohers. This feature is generally not seen in the older, i.e. *sūtra* literature, but in the later philosophical literature, which consists of the commentaries of the *sūtra* literature, the feature is clearly seen. The commentator, generally assumes that the author of the aphorisms or of the basic text was faultless in his exposition. On the occasions where the aphorisms or the basic text which is commented upon cannot be justified in a straight-forward way, the tendency is found to misinterpret the aphorisms/ the basic text. Here the intention is to justify the author. The commentator may not feel that he is misinterpreting because he is doing so with the simple belief that the author cannot commit any mistakes. On one occasion, Jayanta is seen to express this kind of belief in Akṣapāda very clearly when Jayanta says, "A flawful person of short vision like me may talk other things than truth, but the person like Bhagavān Akṣapāda who perceives the far edge of the ocean of logic, cannot speak anything else than truth."[38]

There are some places in the Indian philosophical literature where it is stated that the so-called authority also should not be believed without examination. The distinction is generally made between the statements which cannot be examined either by perception or inference and the statements which can be so examined. *Dharmaśāstra*, the so-called discipline of religious and moral conduct, was supposed to be non-examinable by perception or inference. Therefore it was stated that whatever Manu (the author of *Dharmaśāstra*) says is the medicine. But the discipline of Nyāya, for instance, was a discipline concerned with the phenomena which were subject to examination with the help of perception and inference. Thus Akṣapāda, unlike Manu, was disputed by Nyāya logicians themselves.[39] Vācaspati held that he would not accept any empirically falsified statement, even if it is stressed by thousands of *Vedas*.[40]

However, such statements are rare and they mainly act as the justifications of their thorough respect for the authorities. Whenever it appears to the commentators that the authority seems to be going wrong, the usual trick of misinterpretation is played and the authority is protected (though the original contention of the authority is destroyed by misinterpreting it).

§ 8.44

Ad ignorantiam also does not have any exact Indian parallel. However, in the explanations of non-apprehension *(anupalabdhi)* as a means of knowledge the awareness of such a fallacy is noticed. Ad ignorantiam is described as an appeal to ignorance. One who claims, for instance, that there are no ghosts because no one has ever been able to perceive the existence of any, is committing this fallacy. Indian logicians would hold such an argument as fallacious because according to them, non-existence cannot be proved by mere non-apprehension. The non-apprehension of an object can be used to prove the absence of the object only if the object itself is apprehesible *(dṛśya, yogya, upalabdhilakṣaṇaprāpta)*. The ghost is not a perceptible object. Therefore the fact that the ghost is not perceived does not justify that there are no ghosts.

Another variety of ad ignorantiam is spelled out by an opposite example which runs,

'There must be ghosts, because no one has ever been able to prove that there are not any.'

This may be treated as an example of *anupalabdhisama- jāti*. One who commits this *jāti* argues like the following.

Non-apprehension of the ghosts is not apprehended. Therefore non-apprehension must be absent. Therefore ghosts must be apprehended. Therefore there must be ghosts.[42]

§ 8.45

'Non-cause as cause' is another fallacy which comes under Ignoratio Elenchi. As described by Aristotle, it means a misled application of reductio-ad-impossible. As HAMBLIN describes,

"One way of refuting a thesis is to deduce an obvious falsehood from it and in practice, it will often be the case that the falsehood is deduced not from the thesis alone but only granted certain other assumptions. However, there is a logical trap inherent in the procedure."[43]

As we saw in § 7.32, many of the *jātis* are the misled applications of the method of reductio-ad-impossible. Thus the idea of non-cause as cause comes under these varieties.

§ 8.46

Another fallacy called False cause is different from Aristotle's non-cause as cause.[44] The Latin expression that describes the fallacy of False cause is 'post hoc, ergo propter hoc' which means 'after this, therefore because of this.' Such a faulty argument concerning causation was not distinctly described as a fallacy in ancient Indian logic. But the notion of *anyathāsiddhi* was many times applied to show that the particular event A, though always prior to event B, was not cause of the event B. This notion was systematically used in *Navyanyāya*, but in ancient *Nyāya* also starting from the time of Udyotakara, the notion was effectively used. For instance, in *NMr* Jayanta refers to Mīmāṁsakas arguing that though our use of sense-organs for seeing the place where the pot is absent, is regularly prior to our coming to know that the pot is absent from that place, we cannot therefrom conclude that the sense organs are the means of the cognititon of the absence of the pot. Here the unvitiated temporal order between the use of the sense-organs and the cognition of the absence is, according to Mīmaṁsakas, *Anyathāsiddha*[45]. According to them the cause of the cognition of the absence of a thing, is the absence of the cognition of that thing.

Here *anyathāsiddha* means that which can be explained otherwise. The use of the sense- organs can be explained, in the above case, otherwise, i.e. without bringing into picture the cognition of the absence of the pot. It can be explained as the cause of the cognition of the place where the pot is absent. Therefore while accounting for the cognition of the absence of the pot, the consideration of the use of sense-organs as the means of that cognition would be mistaken.

§ 8.47

The term *anyathāsiddha* is used in a broader context of *vyāpti* in ancient Nyāya. The invariable causal relation to which we referred in § 8.46, can be called a species of the invariable relation called *vyāpti*. Thus, while discussing *anyathāsiddha* we are concerned with the fallacies based on the falsity or the dubitability of the *vyāpti*-statement, i.e. the fallacies of inductive reasoning. *Anyathāsiddha* can be called a case of hasty generalisation when the necessary connection between the two relata is claimed only on the basis of repeated co-existence of them. COHEN and NAGEL analyse a hasty generalisation that 'All red-haired people are ill-tempered' in the following words.

"...the red-haired men we have observed may in fact possess peculiar characters, such as being tired, over-worked, poor and so on, which they do not share with the other members of red-headed group; and their irritability may be due to these peculiar characters".[46]

This is a way of showing that the irritability of red-haired men is *anyathāsiddha*. That is, it can be explained otherwise.

§ 8.48

J.S. Mill divides the inductive fallacies in two main groups

1) Fallacies of observation

2) Fallacies of generalisation[47]

The fallacies of observation are grounded in mal-observation or non-observation. In other words, the arguer making this type of fallacious induction has either not observed or neglected or falsely observed the counter-instance which falsifies the general statement drawn by the arguer. The case of a fallacy of generalisation, however, is a case of generalisation which can neither be falsified by any available counter-instance, nor be justified satisfactorily by the available evidence.

§ 8.49

In Indian logic after Vātsyāyana we have many types of inductive fallacies mentioned. The two varieties of *anaikāntika (sādhāraṇa* and *asādhāraṇa), viruddha* and *anyathāsiddha* are, in fact, some of the types of inductive fallacies. Out of these, the occurrence of *viruddha* in course of argumentation is rare. Secondly in every case of *viruddha*, the essential symptom of *anaikāntika*, namely the demonstratability of *vyabhicāra*-case (= the case in which *hetu* exists but not *sādhya*) is observed. Thus the fallacy of *viruddha* need not be given a special attention. However, other inductive fallacies are worth mentioning. We come across different situations in which the induction is wrong. In following sections § 8.50-54 we shall consider them one by one.

§ 8.50

(I) We may come across a situation where the generalisation can be falsified by pointing out a concrete disconfirming instance where *hetu* exists without *sādhya*. This is seen in the cases of *sādhāraṇa anaikāntika* and *viruddha*. Such a fallacy is closely similar to 'the fallacy of observation' discussed by Mill.

§ 8.51

(II) We may come across a situation where the generalisation can not be falsified by any concrete disconfirming instance but the generalisation also lacks the necessary positive evidence because no confirming positive instance of the general statement is available to us. Here the generalisation is based upon merely negative confirming instances. We find this situation in *asādhāraṇa anaikāntika* as well as in *kevalavyatireki* inference.[48] According to Mill this is a fallacy of generalisation. While describing this kind of fallacious generalisation he says,

"A little reflection indeed will show that mere negations can only form the ground of the lowest and least valuable empirical law. The phenomenon has never been noticed; this only proves that the conditions of that phenomenon have not yet occurred in experience, but does not prove that they may not occur hereafter."[49]

We have seen in § 5.18 that Dharmakīrti is aware of this fallacy which he names as *sandigdhānvayavyatireka.*

§ 8.52

(III) We may come across a situation where the generalisation cannot be falsified by any concrete disconfirming instance but the generalisation also lacks the necessary evidence because no confirming negative instance is available. We find this situation in case of *kevalānvayi-hetu.* As we have seen in § 6.25, this kind of hetu also cannot nullify the interencial risk. We also saw that in the frame of Dharmakīrti's analysis this fallacy is called *sandigdhavyatireka* (See § 5.19).

In western logic, however, I have not found this fallacy discussed.

§ 8.53

(IV) Sometimes, in the case of a usual *anumāna* we may find that the *pakṣa* subject is 'everything.' Now in that case we cannot cite any instances of the *vyāpti*-statement there, because the instances we cite should be external to 'everything.' And there is nothing external to everything. This is the situation of *anupasaṁhārī*, which is not strictly speaking a fallacy of induction. However, we can conceive of a situation where an inductive general proposition cannot have any concrete instanciation. We cannot cite any confirming or disconfirming instance of the general statement. For instance, we consider an artificial statement 'Any horn of a hare does not have the fragrance of a sky-flower.' Here the confirming positive instance should be a horn of a hare, which does not have the fragrance of a sky-flower. But we do not know of any such horn of a hare. The confirming negative instance should be a thing which has the fragrance of a sky-flower, but which is not a horn of a hare. Again we do not know any such thing being there. The disconfirming instance should be a horn of a hare which does have the fragrance of a sky-flower. We are again unaware of any such thing. Such a situation would be a combination of II and III, We do not find it discussed in either Indian or western logic.

§ 8.54

(V) We may come across a situation where the generalisation cannot be falsified by any concrete disconfirming instance, but the generalisation also lacks the necessary evidence because the binding relation *(pratibandha)* cannot be established between *hetu* and *sādhya*. As we saw in § 6..10, Cārvākas claim that any generalisation which is not the result of perfect induction, is bound to be fallacious due to this fallacy.

We see that such a fallacy was called *anyathāsiddha*. Though Dharmakīrti did not recognize this fallacy by this very name, he tried to lay down the conditions in which generalisation can be kept away from this fallacy. For Naiyāyikas, who accepted the method of simple enumeration as a good ground of generalisation, *anyathāsiddha* served as a negative check for confirming the truth of a *vyāpti*-statement. Dharmakīrti on the other hand tried to provide a positive tool for the confirmation of *vyāpti* by fixing the area of *vyāpti*.

Mill discusses a fallacy of generalisation which he calls 'fallacy of mistaking empirical for causal laws.'[50] This fallacy would be identical with *anyathāsiddha* when regarded as a fallacy of causal generalisation. The fallacy occurs when somebody merely uses the method of simple enumeration for establishing a causal law. However, though Mill does not notice, the mere use of the method of enumeration would generate the same difficulties in case of establishing non-causal laws as well. Thus we saw in § 8.47 that *anyathāsiddha* was considered in a broader context of *vyāpti* by ancient Naiyāyikas. Accordingly we can consider 'the fallacy of simple enumeration' which will have a broader scope than Mill's fallacy of mistaking empirical for causal laws.

§ 8.55

As we have seen in § 8.47, COHEN and NAGEL's fallacy of hasty generalisation is similar to *anyathāsiddha*. However, COHEN and NAGEL do not take into account Mill's distinction between the fallacies of observation and the fallacies of generalisation. The example of hasty generalisation they discuss (namely, all red-headed people are irritable) is also a fallacy of observation. Similarly we can claim that *anyathāsiddha* can be pointed out even in the case of *sādhāraṇa anaikāntika* (which is a fallacy of observation). Many times the fallacy is named according to the way in which it is discovered. Thus if a general statement is falsified by pointing out a counter-instance, then the fallacy would be regarded as the fallacy of observation. If the same general statement is refuted on the ground that it is based on simple enumeration, but the necessary relation which is asserted by the statement cannot be established, then the fallacy (in that general statement) is regarded as the fallacy of generalisation.

However we find in Indian logic that *anyathāsiddha* (which is a fallacy of generalisation) is claimed in an inductive reasoning, when a fallacy of observation cannot be shown. As compared to this, the term hasty generalisation is ambiguous because it can connote both, the fallacy of observation and generalisation.

§ 8.56

Thus we can see that the fallacies concerned with *vyāpti*-statement, in Indian logic, can be well compared with the inductive fallacies of the west. Though western logic does not have the notion of *pakṣadharmatā*, the *pakṣadharmatā*-premise of an inference, accepted by Indian logic, would resemble a kind of minor premise of western syllogism. The form of *pakṣadharmatā*-premise would be

'All S is M'. Thus the fallacies based on falsity/dubitability of *pakṣadharmatā*-premise would be compared with the fallacies based on falsity/ dubitability of a minor premise. The types of *asiddha* (excluding *anyathāsiddha)* would come under this group.

§ 8.57

Another important type of material fallacy, as we mentioned in § 8.39, is given to us by 'the fallacies of false or dubitable conclusion.' This type is never found in western logic. Buddhist logicians also, from Dharmakīrti onwards, discarded these fallacies, as these according to them need not be regarded as a distinct type of fallacies. As we saw in § 6.29-31 fallacies of this type, namely *satpratipakṣa* and *bādhita* are only practically important as the distinct ways of refutation.

§ 8.58

We notice some other types of fallacies regarded as important in the context of Indian logic, which are generally neglected by western logicians. We shall consider these one by one.

(I) Fallacies of superfluity : Udyotakara in his long list of fallacies of *hetu* mentions many varieties where a superfluous *hetu* is combined with a good *hetu.*[51] Though the other Nyāya logicians following Udyotakara are not observed to have accounted for these so-called fallacies as *hetvābhāsas*, still superfluity was generally regarded as a fault in debate. We find Śaṅkarasvāmin, a Buddhist logician, mentioning the fallacy of superfluous thesis.[52] Dharmakīrti also accepted that making a superfluous statement in course of debate is an occasion of defeat.[53]

In the west, stoics mentioned the fallacy of superfluous premise. But no other western logicians seem to have taken this fallacy seriously. As pointed out by HAMBLIN, there is nothing in a superfluous premise that needs to be considered to invalidate a proof.[54] We can add to this by saying that mere superfluity of certain premises or even of the conclusion cannot turn a sound argument into an unsound one. However, no Indian logician has explicitly rejected superfluity as a fallacy.

§ 8.59

(II) Contextual fallacies : Starting from Diṅnāga's period we come across several forms of contextual fallacies.[55] These are 'fallacies' only in relation to a particular context of debate. If the context is changed, the same fallacy may turn out to be a good argument. For instance if the disputant while making an argument A uses a *pakṣadharmatā*-premise p, and if p, though true, is not accepted by the opponent, then A is supposed to be fallacious due to *prativādyasiddha*. This is a contextual fallacy; and a question may be raised whether this is a genuine fallacy at all.

Jayanta suggests that *anyatarāsiddha* (= a contextual fallacy in which *pakṣadharmatā*-premise is not accepted by one of the debaters) is not always a fallacy. When a statement of *pakṣadharmatā* stated by the disputant is not accepted as true by the opponent, the disputant has to prove it. If the disputant is unable to prove it, then the disputant is defeated. If, on the contrary the disputant proves it, then the opponent is defeated.[56]

Jayanta's rule suggests that a disputant has committed a genuine contextual fallacy only if he is unable to justify what he has argued. In this way Jayanta tries to remove the drawback of some contextual fallacies. However, in spite of this rule, the contextual fallacies remain contextual fallacies. The disputant's argument may be sound, though he is unable to justify it. In such a situation the disputant may be defeated though he has not committed any fallacy. Thus the distinction between an occasion of defeat and a fallacy is significant here. The 'genuine contextual fallacies' are not fallacies but only the occasions of defeat. The fallacy as such should be non-contextual, i.e., independent of the debate-context. This seems to be the reason why western logicians do not consider contextual fallacies as genuine fallacies.

The same is the case with 'external fallacy of *viruddha*'that we discussed in § 2.14. By generalising this notion we can talk of an external fallacy in general. By external fallacy we mean an argument having a premise or the conclusion, which is contrary to some opinion of the arguer, which is not expressed in the same argument. On the other hand if a premise or the conclusion of an argument is contrary to some step in the argument itself, then it would amount to an internal fallacy. A variety of this fallacy would be *svayaṁ vādino'siddha* (see § 5.16) where the *pakṣadharmatā*-premise of the arguer is negated by his own opinion.

HAMBLIN describes this type of *viruddha* as 'a possibly valid inference open to objection ad hominem.'[57] HAMBLIN's description is misleading because the external fallacy of *viruddha* is not open to any type of objection ad hominem; it is open only to a particular type thereof. Argumentum ad hominem is a fallacious argument addressed to the peculiar circumstances, character, avowed opinions or past conduct of the individual.[58] In the case of external *viruddha*, the argument is open to objection addressed to the avowed opinions of the debater. Though such an objection tells more about the debater than about the argument he is making, the objection has some logical significance, and cannot be hastily labelled as fallacious. When the external fallacy of *viruddha* is discovered in an argument, it raises a doubt regarding the soundness of the argument by pointing out that the premise of the argument must be false if the contrary opinion is true.

§ 8.60

(III) Fallacies of instance : We have seen in § 3.22 that 'instance' as a part *avayava* of argument consists of both the elements of inference : *hetu*-element and *dṛṣṭānta*-element. Due to this, the fallacies of instance were mixed and confused with syntactical fallacies called *ananvaya* etc. We can, however, consider the pure fallacies of instance separately. Fallacies of instance are the fallacies based on falsity or dubitability of the instanciation of the *vyāpti*-statement made by the arguer. It may be seen here that the instanciation of *vyāpti*-statement does not serve any justificatory function. In the frame of Nyāya logic, the function of the instanciation is to show that the *vyāpti*- statement, if true, is true in non-empty universe. In the frame of Buddhist logic, however, the function of the instanciation is an explanatory one. The instanciation indicates the means through which the data for the knowledge of *vyāpti* are apprehended. Thus as the instanciation does not have any justificatory role, the so-called pure fallacies of instance, do not stand for the faults in justification.

On suface-level, the argument which involves a false instanciation is necessarily unsound, because the instanciation being a part of a premise of the argument, such argument does have a false premise. However, the fallacy does not seem to be serious, because if the 'false instanciation' is discarded from the argument, the remaining argument will be sound, provided that the remaining premises are true and the remaining argument is valid. In fact instanciation is not a logically necessary part of a sound argument. Therefore the fallacies of instance need not be taken seriously in the context of valid frame of argument.

The instances have much more importance in the frame of analogical reasoning.

§ 8.61

(IV) Fallacies of analogical reasoning : Though instances do not serve any important purpose in valid reasoning, they have to play a main role in analogical reasoning. In analogical reasoning we pass from the world of observed instances to the world of un-observed thesis. But this is not done on the basis of any universal statement. While making analogical argument, one has to mention therefore, the instance or instances, from which he is passing. Not mentioning any instance, thus, would be a fallacy of analogical reasoning. We see for instance in *Carakasaṁhitā*, that the fallacies called *prakaraṇasama* and *saṁśayasama* were the arguments which lack the mention of any instance. (See § 1.6) Diṅnāga also would regard such an absence of instance as a logical fault called *nyūna*. (See § 5.9). We also saw that the concept of analogical reasoning underwent a change in the hands of Caraka, Akṣapāda, Vātsyāyana, Diṅnāga and and Udyotakara. In this process of history we have the concepts of stronger and stronger analogical reasoning. As the notion of analogical reasoning changed, the notion of fallacies of analogical reasoning changed. After Caraka, the observed general statement was regarded as the necessary part of analogical reasoning and for Diṅnāga, to mention only an instance without 'generalisation in the observed world', was a fallacy of instance (See § 3.39-40). The fallacies of instance which we mentioned in § 8.60 also are significant as the fallacies of analogical reasoning.

In fact the history of fallacies discussed in pre-Dharmakīrti period, was the history of fallacies of analogical reasoning. As we saw in § 2.3, inductive reasoning is a modification of analogical reasoning. Therefore, many fallacies which were later on regarded as inductive fallacies (i.e. the fallacies based on falsity/dubitability of the *vyāpti*-statement) were presented in these days in the form of the fallacies of analogical reasoning. Further it may be seen that during the period of Akṣpāda and Vātsyāyana, when the 'analogical reasoning' was presented in a weaker form, the inductive /analogical fallacies of observation,[59] like *sādhāraṇa anaikāntika* and *pratijñāvirodha*[60] were formulated. But during this period the inductive fallacies of generalisation were not taken note of. When the analogical reasoning was discussed in a stronger form in the hands of Diṅnāga, the inductive/analogical fallacies of generalisation like *asādhāraṇa* were also taken note of, but the more subtle fallacy of generalisation namely *anyathāsiddha* was first located by Udyotakara.

Fallacies of analogical reasoning are thus significant in the context of pre-Dharmakīrti Indian logic. When from the period of Dharmakirti the form of inference was conceived of as a valid form, some fallacies of analogical reasoning were converted into inductive fallacies. But since the mention of an instance in the argument was not regarded as that significant (as before), the pure fallacies of instance gradually disappeared from the scene.

Notes

1. बुद्धिः पश्यति या भावान् बहुकारणयोगजान् ।
   युक्तिस्त्रिकाला सा ज्ञेया त्रिवर्गः साध्यते यया ।। *CS*, p. 155.
2. See *PV, svārthānumāna*, verse 31.
3. See *Ibid*, verses 34 to 37.
4. तस्माद् दृष्टस्य भावस्य दृष्ट एवाखिलो गुणः ।
   भ्रान्तेर्निश्चीयते नेति साधनं सम्प्रवर्तते ।। *Ibid*, verse 45.
5. Although Siddhasena, the Jaina logician who preceded Akalaṅka classified *pramāṇa* into *Pratyakṣa* and *parokṣa* he in his *Nyāyāvatāra* almost identified *parokṣa* with *anumāna*. Akalaṅka therefore seems to be the first Jaina logician to classify *Parokṣapramāṇa* into *smṛti, pratyabhijñā, tarka, anumāna* and *āgama*. For the discussion of *tarka* and Akalaṅka's conception of it, see *Akalaṅka's Criticism of Dharmakīrti's Philosophy, A study*, by Dr. Nagin J. Shah, L.D. Institute of Indology, Ahmedabad (1967), pp. 259-264.
6. Jayantabhaṭṭa, for instance, places *tarka* in between *pramāṇa* (that gives knowledge with certainty) and *saṁśaya* (doubt) and describes the knowledge based on *tarka* as probabilistic cognition which is of middle or third kind *(madhyavartinī tṛtīyā saṁbhāvanāpratītiḥ* (see *NMr*, Vol, II, p. 145). Akalaṅka too agrees with the probabilistic character of *tarka*. He describes *tarka* as *saṁbhavapratyayaḥ* (See Dr. Nagin J. Shah, op. cit., p.262, fn. 41)
7. See *ILSM*, p. 275.
8. For the discussion of perfect induction and demonstrative inductive reasoning see *ILSM*, pp. 275-278.
9. See *MIIL*, pp. 156-157.
10. See *PBTL*, Introduction.
11. See §§ 6.31, 32.
12. Vātsyāyana defines *hetvābhāsa* as something which appears as a *hetu* proper, but actually is not.

    हेतुलक्षणाभावाद् अहेतवो हेतुसामान्याद् हेतुवदाभासमानास्त इमे ..... *NBh* as the introduction to *NS* 1.2.4.
13. See *FAL*, p. 12.
14. See for instance, *MIIL*, p. 154, *NTK* Chap. XIV.
15. The dialogue is based on an example of amphiboly mentioned in *FAL*, p. 16.

16. The author of *Upāyahṛdaya* has included *chala* in *hetvābhāsas*. See *Upāyahṛdaya* (included in *PBTL*) p. 14.

17. हेत्वाभासाश्रया खल्वियम् अव्यवस्था । *NBh* under *NS* 5.1.3.

18. See *FAL*, pp. 169-70.
See § 6.32

19a. Here I am taking *ananvaya* and *apradarśitānvaya* as synonyms. (Similarly, *avyatireka* and *apradarśitavyatireka*). If following Dharmakīrti we make a sharp distinction between the two (see § 5.28), then we have to regard *ananvaya* and *avyatireka* as the combinations of a semantical fallacy and a syntactical fallacy. Thus *ananvaya* or *avyatireka* would not be one fallacy but two fallacies, whereas *apradarśitānvaya* and *apradarśitavyatireka* would be purely syntactical fallacies.

20. See *TS*, p. 318. CHATTERJEE holds the same view.
See *NTK*, p. 281.

21. See *TS*, pp. 321-6.

21a. Here we have avoided the discussion of the issue whether I or O propositions can be converted into the proper theses of Indian *Parāthānumāna*. However, I have a feeling that such an attempt can be made. See Appendix IV.

22. See *TS*, p. 325.

23. Ibid

23a. Indian logicians were aware of the role of accent in determining the meanings of sentences uttered. In recitation of *Saṁhitā*-texts of *Vedas*, the accents on the vovels were fixed and any deviation was prohibited not only from religious point of view, but from semantical point of view as well. An ancient myth goes according to which, a demon called Vṛtra who wanted to kill Indra through his son, found his son killed by Indra because his prayer as 'Oh Vrtra's son! Grow and become *Indraśatru*' (where Indraśatru was supposed to mean 'the killer of Indra') meant a complotely opposite thing, as 'Oh Vṛtra's son ! Grow and let Indra be your killer.' This happened only because the word *Indraśatru* in the prayer was uttered with a different accent. The myth suggests the ancient Indian awareness of the fallacy of accent. (The story is referred to for example in *VMB*, p. 19).

24. See FAL, pp. 25-6; GL, p. 342

25. Amongst the modern scholars of Indian logic, BARLINGAY seems to be the first to notice this distinction implicit in ancient Nyāya and Vedānta treatises. In his article on G.E.MOORE, written in 1957-58, he says, "The western writers do talk of part and whole,

but they have not distinguished between wholes and parts which are homogeneous and wholes and parts which have a heterogeneous relation. The table, for example, has a heterogeneous relation with its parts. A leg of a table is not a small table. Nor is table a big leg (of a table). But a drop of water has the homogeneous relation with the expanse of water. Indian writers have distinguished two kinds of wholes. Where there is homogeneous relation, they term as Aṁśa-Aṁśi-relation, where there is heterogeneous relation, they term as Avayava-Avayavi- relation." The article is now published in ***Beliefs, Reasons and Reflections***, 1983.

26. See *FAL* p. 213.
27. See *FAL*, p. 211.
28. The Nyāya logicians would not go by this way. They would try to remove the ambiguity by using the expressions like *avaccheda* (= limit). A popular Navya-Nyāya example runs -

    1) वृक्षः कपिसंयोगी ।(= The tree has a monkey on it.)

    2) वृक्षः कपिसंयोगाभाववान् । (= The tree does not have any monkey on it.)

The apparent contradiction will be removed by interpreting the two statements as

1) शाखावच्छेदेन वृक्षः कपिसंयोगी । (As far as the branches are concerned, the tree has got the monkey.)

2) मूलावच्छेदेन वृक्षः कपिसंयोगाभाववान् । (As far as the root is concerned, the tree does not have any monkey.)

This expresses the Navya Nyāya awareness of the Boethian fallacy of different part. In ancient Nyāya also the words like avaccheda are used effectively, but not as frequently and keenly as in Navya Nyāya.

29. See *FAL*, p. 171.
30. See § 2.16.
31. See § 7.74.
32. See *FAL*, p. 31.
33. See Whately's tree of fallacies, *FAL*, p. 171.
34. See *FAL*, p. 41
35. See *FAL*, p. 42.
36. See *NMr*, Pt. I, p. 138.
37. See *FAL*, p. 42.
38. See Note 7 under Chap. VI.
39. See for instance, *NMr*, Pt. I, p. 74.
40. न ह्यागमाः सहस्रमपि घटं पटयितुम् ईशते । *BHAM*, p. 3.

41. Compare this with the form of the *jāti* given in § 7.27.
42. See § 1.2.
43. *FAL*, p. 78.
44. See *FAL*, p. 38.
45. तत्र यथा रूपानुमीयमानस्पर्शविदने नयनान्वयव्यतिरेकान्वयविधानम् अन्यथासिद्धम् एवम् इहापि भूप्रदेशोपलम्भाविनाभाविनि कुम्भाभावग्रहणे तत्कृतम् इन्द्रियान्वयव्यतिरेकान्वयविधानम् इति न चाक्षुषो घटाभावप्रतिभासः । *NMr*, Pt. I, p. 47.
46. *ILSM*, p. 277-8.
47. See *SLRI*, Book V, Chapter II.
48. See § 6.24.
49. *SLRI*, p. 789.
50. See *SLRI*, p. 788
51. See § 4.15.
52. See § 3.19.
53. See Dharmakīrti's second basic type of occasion of defeat as stated in § 7.57.
54. See *FAL*, p. 215.
55. The principle of acceptance is the governing principle of such fallacies. See §§ 3.25, 4.12, 5.16, 6.30, 7.79.
56. See *NMr*, Pt. II, p. 163.
57. See *FAL*, p. 181.
58. See Whately quoted by HAMBLIN, *FAL*, P. 174.
59. See § 8.48.
60. *Pratijñāvirodha* is identical with the internal fallacy of *viruddha*. See § 7.37.

Chapter IX

Conclusions

§ 9.0

In the following sections of this chapter I wish to narrate the major observations we arrived at in the previous chapters. I would classify the conclusions in three groups.

(1) Conclusions in the history of philosophical ideas. (In this section I would narrate the historical observations regarding ancient Indian concepts of inference, fallacy and debate.)

(2) Conclusions in philosophical logic. (In this section I would narrate the observations regarding the nature and logical status of the ancient Indian concepts of inference, fallacy and debate.)

(3) Conclusions in comparative philosophy. (In this section I would narrate the observations about the comparison between western and ancient Indian awareness of inference and fallacies.)

While narrating the conclusions I am aware that the clear demarcation-line between the three, especially between the first and the second, cannot be drawn. My own observations about the nature and status of the concepts of inference, fallacy and debate, are obviously included in the second group. But the observations about the respective positions of the ancient Indian logicians can be included in both, first and second group. In the first group I have included those observations which I felt historically more significant, though they may be philosophically significant as well. In the second group I have included those observations which I felt philosophically more significant, though they may be historicall, significant as well.

The bracketed roman numbers refer to the numbers of the chapters from which the observations are singled out.

§ 9.1

Conclusions in the History of Philosophical Ideas.

On Inference: Caraka's account of *anumāna* was ambivalent. He could not harmonize the scientific aspect of universal causal laws with the traditional aspect of analogical reasoning which he borrowed probably from Medhātithi Gautama. The scientific insight shown by Caraka in presenting the notion of *yukti* was not pursued by the successive Nyāya logicians(I).

When for Caraka, inference was mainly a speculation about some unobservable phenomena, for Akṣapāda, a major class of *anumānas* consists of those *anumānas*, which have an empirically verifiable conclusion (II).

While defining *hetvābhāsa* Vātsyāyana talks of 'defining characteristics' (*lakṣaṇa*) of good *hetu*, but he does not explain the 'defining characteristics' or CEs of *hetu* (II).

Diṅnāga was the first logician to work out the theory of reasoning based on three CEs of *hetu*. He was the first to distinguish clearly between *svārthānumāna* and *parārthānumāna*. Diṅnāga suggested the possibility of *kevalānvayi* inference, which was purely positive analogical reasoning (III). Udyotakara who came after Diṅnāga, however, explicitly mentioned the three types of analogical reasoning: purely positive, purely negative and positive- negative (IV).

Diṅnāga talked in terms of some regular relation between *hetu* and *sādhya*. But he must not have thought in terms of necessary relation. Dharmakīrti misinterprets Diṅnāga, when Dharmakīrti claims that Diṅnāga's use of the term *niścita* indicates his interest in necessary relation (III).

It has been stated by many scholars that the concept of pervasion occurs first in the treatise of Udyotakara. But in fact Udyotakara's (so-called) conception of *vyāpti* is restricted to observed world. Diṅnāga's basic frame of inference as strong analogical reasoning was not given up by Udyotakara (IV). The early logicians like Vātsyāyana, Diṅnāga and Udyotakara use the term *dṛṣṭaṁ* in the examples of the seemingly general *'vyāpti'*- statements. This indicates that their *vyāpti* was limited to the observed world and was not the universal *vyāpti* as was later on conceived of (II - IV).

Of course the concept of *anumāna* as analogical reasoning seems to have assumed stronger and stronger form in the course of history from Caraka to Udyotakara (I - IV).

Dharmakīrti was the first to consider *vyāpti* as universal and necessary relation between *hetu* and *sādhya*. His special use of the terms *siddha*, *dṛṣṭa*, *sarva*, *eva* and *asambhava* indicates this (V). It can thus be said that Diṅnāga's concept of inference and Dharmakīrti's concept of inference were basically different from each other.

In Indian logic upto Udyotakara, the practical aspect of reliable reasoning was emphasised, but the logical aspect was ignored, whereas Dharmakīrti emphasised the logical aspect without ignoring the practical aspect (VI).

In general it may be said that *anumāna* according to Caraka, Akṣapāda, Vātsyāyana, Dinnāga and Udyotakara was of the type of pre-inductive-generalisation reasoning whereas Dharmakīrti onwards it presupposed inductive generalisation and thus was conceived as post-inductive inference (II - VI). It is also seen that as there was development in the account of *anumāna*, there was also a similar development in the critique of *anumāna* (VI).

The Nyāya logicians starting from Jayanta accepted the notion of *vyāpti* as a universal and necessary relation, but still accepted the five elements as the necessary CEs of *hetu* (VI).

In ancient Nyāya logic the extralogical considerations are seen to have entered and poluted the logical problem of *kevalānvayi* and *kevalavyatireki* inference (VI).

On fallacies: Caraka was aware of the concepts of *chala, jāti (uttara), hetvābhāsa (ahetu)* and *nigrahasthāna*. Akṣapāda's account of the same resembles more in form and content with the account of Caraka than with the developed account of the later logicians like Jayanta and Vācaspatimiśr a(I).

While giving examples of *vyabhicāra*, Vātsyāyana gives an example, which is not an example of *vyabhicāra* at all (§ 2.12). This attitude suggests that Vātsyāyana had not properly worked out the concept of *hetu-sādhya*-relation. Even Udyotakara was not aware of Vātsyāyana's mistake, and although Vācaspati was aware of it, he only tried to protect him by misinterpreting him(II).

The *viruddha* of Akṣapāda and Vātsyāyana is different from the popular *viruddha* of Diṅnāga onwards. But the later Naiyāyikas like Jayanta identified the two by misinterpreting Akṣapāda's definition of *viruddha* (II). *Prakaraṇasama* of Akṣapāda and Vātsyāyana closely resembles the one of Caraka (II). While discussing *hetvābhāsas* Diṅnāga uses the principle of acceptance along with the principles of falsehood and dubitability (III). This trend was followed by all the later Indian logicians.

Diṅnāga while giving the list of fallacies does not include the cases of non-existence of *sapakṣa*, *vipakṣa* or both. Later on Udyotakara considers these (III, IV).

Udyotakara is more keen than Diṅnāga as far as he conceives of the different subtle factors which make a *hetu* fallacious. But Diṅnāga had anticipated some of the elementary defective states of *hetu* referred to by Udyotakara (IV).

Udyotakara's account of *anyathāsiddha* makes us think of some notion of 'necessary relation' between *hetu* and *sādhya*. But such a notion is not positively noticed in Udyotakara's writings (IV).

Diṅnāga's *nyūna* was not basically a formal fallacy whereas Dharmakīrti's *nyūna* was. Dharmakīrti explained *asādhāraṇa-anaikāntika* in a different manner from one in which Diṅnāga did (V).

Dharmakīrti in *Vādanyāya* refined and systematised Akṣapāda's list of the 'occasions of defeat' (*nigrahasthānas*). He did deny that *aprāptakāla* is a proper occasion of defeat and allowed *nyūna* as an occasion of defeat only in a restricted sense. (VII)

Jayanta identified *satpratipakṣa* with *prakaraṇasama* of Akṣapāda and all the later Naiyāyikas follow Jayanta (VI).

In the early theory of *anumāna*, the fallacy called *savyabhicāra* functioned as a negative check upon the strength of analogical reasoning, but the mention of *savyabhicāra* as a fallacy did not give *anumāna* the status of post-inductive inference. From Dharmakīrti onwards when *anumāna* took the shape of post- inductive reasoning, *savyabhicāra*, which was earlier a fallacy of analogical reasoning, assumed the shape of an inductive fallacy, possible in the case of post-inductive reasoning. In fact all the fallacies discussed in pre-Dharmakīrti period were the fallacies of analogical reasoning. The fallacies of inductive generalisation were the products of evolution in the earlier concept of the fallacies of analogical reasoning (II, VIII).

In general it can be said that the Nyāya-Buddhist awareness of fallacies underwent a change along with the change in Nyāya-Buddhist concept of inference (I-VI).

There is a distinction between *hetvābhāsa* and *anapadeśa* (*aliṅga*) which point respectively to the appearance-aspect and unsoundness-aspect of fallacies. Before Diṅnāga the first, while from Diṅnāga onwards, the second aspect was emphasised.

On the Method of Debate: Caraka prescribes to the physician debaters, not only the intellectual non-academic ways of defeating the opponents, but also the non-intellectual ways of defeating the opponents. The non-intellectural ways, which have psychological bearing only, are not found as prescribed in the later treatises on the method of debate (I).

§ 9.2

Conclusions in Philosophical Logic

On Inference: Vātsyāyana's 'general statement' was a statement of 'observed equivalence' and because of this notion Vātsyāyana misconceived the fallacy called *savyabhicāra* (II). Any so-called inductive reasoning is a modified form of analogical reasoning. Empirical generalisation arises from the combination of observation and analogy. Thus *vyabhicāra* is the fallacy of both, induction and analogical reasoning (II).

Even all the assertions suggested by the three Dinnāgian CEs of *hetu* taken together, do not provide the logically sufficient ground for the proof of *pakṣa*. Dinnāga's form of *parārthānumāna* may be regarded only as a form of analogical reasoning (III).

The necessary relation between *hetu* and *sādhya* according to Dharmakīrti, need not always be logically or conceptually necessary relation. For him, even the *vyāpti*-statement based on *tādātmya* is not an analytic statement (V). Dharmakīrti distinguishes between *sāmarthya* and *pratibandha*. This shows that he was perhaps aware of some distinction between material implication and entailment (V).

Dharmakīrti regards the statement of *pakṣa* to be redundant for *parārthānumāna*. However he is not fully opposed to the importance of *pakṣa* in the context of *parārthānumāna* (V). In fact the statement of *pakṣa* in the form of *pratijñā* or *nigamana* is necessary while making an argument because many conclusions can follow from one and the same set of premises and the thesis which a debater is intending to prove from the premises he is stating, may not become clear to another debater unless it is verbally stated by the arguing debater (VII).

In every analogical reasoning there is a risk. In fact we cannot deny the risk involved in any inferring, although we cannot help inferring. Experience should, thus, not only be regarded as the basis of all knowledge, but also the justificatory principle of all knowledge (VI).

Even the five CEs of *hetu* suggested by Naiyāyikas are not the sufficient conditions of sound *hetu* (VI).

Most of the ancient Indian philosophers in fact believed in some kind of *svataḥ prāmāṇya* though they are actually divided in two groups, *svatah prāmāṇyavādin* and *parataḥ prāmāṇyavādin* (VI).

Bhāsarvajña answers the problem of superfluity of inference but he or any Indian logician does not seem to have answered the problem of warrantable truth of *vyāpti*, satisfactorily (VI).

In a sense purely negative sound difference is impossible. The so-called *kevalavyatireki* inference cannot be clearly distinguished from the inference which is a case of *asādhāraṇa-anaikāntika* (VI).

Cārvākas expressed important doubts which show that any *anumāna* could be basically fallacious (VI).

On Fallacies: Caraka's *prakaraṇasama* and *saṁśayasama* are two varieties of begging the question. In both of them the method of analogy is not applied at all even though it should have been applied. Caraka's *varṇyasama* is in fact a fallacy of instance (I).

Prakaraṇasama of Akṣapāda and Vātstāyana does not resemble petitio principii as RANDLE thinks. MATILAL is right in thinking that *Sādhyasama* of Akṣapāda and Vātstāyana is not the same as petitio principii. *Sādhyasama* closely resembles *sandigdhāsiddha* of Diṅnāga and Dharmakīrti (II).

Kālātīta-hetvābhāsa of Akṣapāda and Vātsyāyana is not only different from *aprāptakāla* but also from *bādhita*. *Kālātīta* arises due to the confusion of time-factor involved in the occurrence or appearance of *hetu, sādhya, pakṣa* or relation between them. Udyotakara's distinction between *upalabdhikāla* and *saṁyogakāla* is significant in this context (II).

According to Diṅnāga, inference consists of three elements namely *pakṣa, hetu* and *dṛṣṭānta*. Diṅnāga discusses the fallacies of each element, under the names *pakṣābhāsa, hetvābhāsa* and *dṛṣṭāntābhāsa* respectively. Diṅnāga's account of fallacies is mainly based on the principles of falsefood, possibility of falsehood and non-acceptance. Out of these basic principles, *pakṣābhāsa* in based only on falsehood (III).

It is rather difficult to demarcate clearly between *hetu*-element and *dṛṣṭānta*-element of analogical reasoning. Thus we find that some fallacies of *hetu* (namely the syntactical fallacies) are accounted under *dṛṣṭāntābhāsa* (III).

The fallacies which occur due to non-acceptance rather than falsehood or dubitability are contextual fallacies and in this sense, relative fallacies (III).

Generally we might suppose that if a *hetu* possesses all the three (Dinnāgian) CEs of *hetu* then the inference from that *hetu* must be sound. But the fallacy called *viruddhāvyabhicāri* may occur in spite of the *hetu's* possession of all the Dinnagian CEs (III).

Udyotakara uses the method of combination on a large scale and counts in all 2032 combinations, which correspond to 2032 types of *hetu* out of which only five refer to the types of sound *hetu* and the rest to the fallacies of *hetu*. The fallacies he discusses are based on the principles of falsity, dubitability, non-acceptance and superfluity (IV).

Dharmakīrti gives greater importance to *hetu*, than to *pakṣa* or *dṛṣṭānta* and discusses *hetvābhāsas* in great details (V).

Dharmakīrti's fallacy in which *vyatireka* (and not *anvaya*) is doubted, is similar to *kevalānvayi* type of inference (V).

Dharmakīrti's supposition that the intended *sādhya* need not be distinguished from the stated *sādhya* seems to be wrong, because we can discover a fallacy in a particular argument only by examining the explicit linguistic form and content of the argument (V).

As Dharmakīrti's form of inference is a valid one, *virudhāvyabhicāri* and *bādhita* for him need not be regarded as distinct fallacies but are reducible to the cases of *savyabhicāra* (V).

The fallacies of conclusion are logically reducible to the fallacies of false or dubitable premises or to formal fallacies. But they still have a separate practical import (IV).

Although *anyathāsidha* is traditionally regarded as a type of *asiddha*, in fact it is not a type of *asiddha*, but is based on dubitability of *vyāpti*-statement (VI).

All the types of *chala* are the types of fallacious refutations and in this sense they can be regarded as the types of *jāti*. Thus there is no water-tight compartment between *chala* and *jāti*. The only difference seems to be that *chala* is always based on deliberate misinterpretation of the opponent's sentence, whereas *jāti* need not be necessarily so (VII).

It cannot be always the case that *jāti*-argument is a formal fallacy. It can even be just a material fallacy or just semilogical fallacy when it involves quibbling (VII). Basically *jātis* are semantical fallacies (VII). Many *Jātis* refer to improper uses of the reductio ad absurdum method (VII).

Prasaṅga-fallacies are shown by indirect method of disproof. We can distinguish between existential *prasaṅga*-fallacies and the epistemic *prasaṅga*-fallacies. While considering the fallacy of epistemic circular dependence, we have to distinguish between epistemology of cognition and epistemology of justification (VII).

Infinite regress (*anavasthā*) may not always be a fallacy (VII).

Nayābhāsas can be regarded as the fallacies of description. Jainas seem to claim that our ordinary language discourse allows the use of ambiguous sentences. But Jaina philosophers have not seriously tackled the problem of removing ambiguities. Secondly many a time the metaphysical inquiry of Jainas and their view of language do not go hand in hand (VII).

Generally *avyāpti, ativyāpti* and *asambhava* are regarded as the fallacies of definition. But the philosophers like Jayarāśibhaṭṭa have rightly referred to 'mutual dependence' as a fallacy of definition (VII).

The Nyāya attitude of bringing all fallacies under *hetvābhāsa* is not only orthodox, but misleading. The term *sādhanābhāsa* appears to be the proper term for fallacy (VIII).

Not only *hetvābhāsa, pakṣābhāsa* and *dṛṣṭāntābhāsa*, but also *chala, jāti, nigrahasthāna* (partly), *prasaṅga*-fallacies, *nayābhāsa* and the fallacies of definition can be regarded as the types of *sādhanābhāsa* (VIII).

All the fallacies which come under *hetvābhāsa, chala, nayābhāsa, nigrahasthāna,* fallacies of definition and *prasaṅga* are semantical fallacies, while *nyūna* (as accepted by Dharmakīrti), *ananvaya, avyatireka, viparītānya* and *viparītavyatireka* are all syntactical fallacies. In fact all the formal fallacies can be included in *nyūna;*

ananvya, viparītānvaya etc. can be regarded as the types of *nyūna* (VIII).

On the Method of Debate: When we discuss hostile debate we have to distinguish between academic and non-academic ways of winning a controversy. The non-academic ways can again be classified into two classes namely intellectual and non-intellectural (I)

In ancient Indian logic, hostile debate was described like a serious game. We can distinguish between basic rules and flexible rules of hostile debate and correspondingly between categorically proper occasions of defeat and conditionally proper occasions of defeat. All *nigrahastānas* are not fallacies, nor are they even categorically proper occasions of defeat. Some *nigrahasthānas* (like *aprāptakāla*) are neither categorically nor conditionally proper occasions of defeat though they were conventionally regarded as occasions of defeat. (VII).

§ 9.3

Conclusions in Comparative Philosophy

On Inference: Dilemma of inference as presented by Cārvākas resembles the one presented by western logicians (VI).

The Indian conception of *anumāna* and the western conception of inference are distinct in some important respects.

We do not find Indian logicians classifying inference into inductive and deductive as we find western ones. But Caraka and Jaina logicians from Akalaṅka onwards are seen to stress the reasoning-aspect of inductive generalisation and regard it to be at par with *anumāna*. Dharmakīrti and the post-Udyotakara Naiyāyikas, however, assimilate induction with intuitive perception rather than with inference for some important philosophical reasons. (VIII).

The ancient classification of *anumāna* into *dṛṣṭa* and *sāmānyatodṛṣṭa* can be compared with the one between scientific and metaphysical reasoning as it is understood in the west. *Pūrvavat* and *śeṣavat anumānas* are exercises with our belief in temporal and spatial aspects of homogeneity of Nature respectively. The *vyāpti*-statements belonging to *svabhāvānumāna* and *kāryānumāna* are comparable with the statements expressing analytic and synthetic necessity respectively as they are understood in the west.

However, the goal of *anumāna* in Indian logic unlike the goal of deductive reasoning in western logic, is truth and not formal validity. Formal validity can at best be one of the necessary conditions to the inferential knowledge of a reality; but it is not an end in itself (VIII).

On Fallacies: Indian logicians are more interested in material fallacies than in formal fallacies. But the charge against the Indian logicians that none of them has discussed any formal fallacies is plainly wrong (VIII).

Any case of 'illicit process of minor' is the case of '*nyūna* due to the lack of *pakṣadharmatā*-statement'. But corresponding to other formal fallacies discussed in the West, we do not have exact parallels in Indian logic. Of course generally it is possible to identify any example of a 'western' formal fallacy as an example of some 'Indian' formal fallacy (VIII).

The fallacies called 'Affirming the consequent' and 'Denying the anticident' look like 'viparītānvaya' and *viparītavyatireka* respectively, although the difference remains. (VIII).

The fallacies like Accident and Secundum quid are convertible into *chala* or *nayābhāsa*. Amphiboly, Equivocation and Accent can be regarded as the subtypes of *vākchala* (VIII).

Figure of speech in one sense resembles *upacāracchala*, in another sense *Śabdanayābhāsa*, *evambhutābhāsa* and *sambhirūḍhābhāsa*. The awareness of Indian philosophers about the fallacy of Composition and Division is expressed through the Nyāya-Vedānta distinction between *aṁśī* and *avayavī* (VIII).

The so-called Boethian fallacies involve the same principle as the *nayābhāsas* do (VIII). Jainas seem to claim that the philosophers who are misled by ordinary language, and are engaged in context-free interpretations and implications of the ordinary language descriptions, commit *nayābhāsas*. This position of Jainas resembles the Rylian position about systematically misleading expressions. While discussing the seven *nayābhāsas* Jainas defend the common-sense view of the world against the so-called philosophical views presented by various philosophers (of the schools other than Jainism). But Jainas cannot be called common sense philosophers in the same sense in which G.E. Moore is one (VII).

Asiddha, viruddha and *savyabhicāra* do not have the exact parallels in western logic, though they can be included in 'fallacies of premise unduly assumed' (VIII).

Prakaraṇasama is similar to petitio principii. But the *prasaṅga*-fallacies like *itaretarāśraya* need to be distinguished from petitio principii. *Anavasthā* and infinite regress are identical (VIII).

The essence of Ignoratio elenchi is clearly seen in various types of *Jāti*. The Indian awareness of Ad ignorantiam is noticed through the ancient Indian account of *anupalabdhi* and *anupalabdhisamā jāti* (VIII).

False cause is similar to *anyathāsiddhi*. Falsity of *vyapti*-statement causes inductive fallacies. Indians seem to have been aware of the inductive fallacies of observation (e.g. *sādhāraṇa anaikāntika, viruddha*) and inductive fallacies of generalisation (e.g. *asādhāraṇa anaikāntika, anyathāsiddha*) (VIII).

Committing Ad Verecundiam is observed to be a salient feature of traditional Indian philosophers. Thus they would not regard it as a fallacy. On the other hand some 'fallacies' seriously discussed in Indian logic, e.g. fallacies of false/dubitable conclusion, fallacies of superfluity, contextual fallacies, 'external' fallacies, fallacies of instance, fallacies of analogical reasoning are not treated as fallacies in western logic (VIII).

APPENDIX I

Dates of the Indian logicians dealt with in the work

1.	Caraka	About 78 (V)
2.	Umāsvāti	1-85 (V)
3.	Akṣapāda	About 150 (V)
4.	Vātsyāyana	450-500 (P)
5.	Diṅnāga	Circa 450- 520 (V)
6.	Praśastapāda	550-600 (P)
7.	Śaṅkarasvāmin	About 550 (V)
8.	Udyotakara	550-610 (P)
9.	Samantabhadra	600 (V)
10.	Dharmakīrti	About 635-650
11.	Jayarāśibhaṭṭa	600-650 (S)
12.	Akalaṅkadeva	750 (V)
13.	Jayantabhaṭṭa	840-900 (P)
14.	Dharmottara	847 (V)
15.	Bhāsarvajña	860-920 (P)
16.	Vācaspatimiśra	900-980 (P)
17.	Udayana	1050-1100 (P)
18.	Vādidevasūri (or Devasūri)	1086-1169 (V)

Note: 1. Bracketed V refers to 'the date accepted by VIDYĀBHUṢAṆA' (See accounts of the respective logicians in *HIL*) 2. Bracketed P refers to 'the date accepted by POTTER' (See *EIP*, pp. 9-10) 3. Bracketed S refers to 'the date accepted by SANGHAVI and PARIKH' (See Introduction to *TUS*, p.x.)

APPENDIX III

Symbolic presentation of the forms of *parārthānumāna.*

1. Vātsyāyana's Notion of *Sādhyasādhanabhāva.*

We have seen in the second chapter that Vātsyāyana's concept of the constant relation between *hetu* and *sādhya* is that of 'observed equivalence'. We can symbolize this form of *sādhya-sādhanabhāva* as.

$\{ (x) (Hx \equiv Sx) \}_{0x}$

Here by Hx I mean 'the substratum x has *hetu* as its superstratum'; by Sx I mean 'the substratum x has *sādhya* as its superstratum'; and by $\{ \}_{0x}$ I mean 'in the realm of the observed values and properties of x'. The above stated form is obviously equivalent with

$\{ (x)[(Hx . Sx) \vee (\sim Hx . \sim Sx)]\}_{0x}$

and with $\{ (x) [(Hx \supset Sx). (Sx \supset Hx)]\}_{0x}$

The case of positive instance indicates those values of x for which Hx.Sx is true and the cases of negative instance indicate those values of x for which ~Hx. ~Sx is true. In such a frame of *anumāna*, any observed value of x for which either Hx. ~Sx is true or Sx. ~Hx is true would give a case of *vyabhicāra.*

2. Diṅnāga's Forms of *Parārthānumāna.*

Let us assume that 'a' stands for the subject of the thesis and 'S' stands for the predicate of the thesis. That is to say, S is the predicate that is to be demonstrated to be predicated of the subject a.

Thus, Sa stands for our *pakṣa* (thesis).

Let us assume that 'H' stands for the *hetu* which is again a predicate. Now as we have seen in § 3.2, the first CE states that hetu is predicated of the *pakṣa*-subject; in other words, that H is predicated of a.

(A) Thus the first CE can be symbolized as Ha.

supposed to be the superstrata and the particular objects as the substrata. But Buddhist logicians generally do not prefer this form which presupposes the existence of universals. They prefer the form 'yat .... tat .... ' which suggests that *vyāpti* ultimately should be reduced to a relation amongst the particulars. Of course *vyāpti* for them is not concerned with only some particulars but it is concerned with all particulars (observed or apriorily known). Thus the Buddhist form of *vyāpti* like 'Whatever is real is impermanent' needs to be distinguished from the Nyāya form of *vyāpti* like 'Wherever there is smoke there is fire'.

Like the term 'universality' the term 'class' is also misleading. 'Class' may mean a set, a collection of things or may mean a common head under which various things are brought, which share something in common. The first sense seems to be nearer to the Buddhist concept of *sāmānya* (which is not *jāti*) and the second sense is nearer to the Nyāya concept of *sāmānya* (which is almost a synonym of *jāti*).

Whatever has been said above indicates as to how Nyāya concept of universality differs from the Buddhist concept. The difference pertains to the metaphysics of universality. But it would be a further question as to whether, and if yes, how this difference in the metaphysics of universality would affect the logic of universality. Metaphysically 'Whatever has .... has ....' and 'Wherever .... there ....' seem to present one model namely *ādhāra-ādheya*-model, while 'whatever is .... is ....' seems to present another model, which may be, for the sake of convenience, called the subject-predicate model. In both the models there is a tendency to 'divide' things into conceptual parts. But the difference is that in *ādhāra-ādheya*-model, the 'division' is more or less confused with physical division, while in the case of subject- predicate model, the 'divison' may be understood to be merely epistemic or linguistic. In logic we take subject- predicate-model as the official model and we try to translate for our logical purposes all statements into subject- predicate language. I feel that in spite of the metaphyscial overtones, a universal statement in *ādhāra-ādheya*-language can be translated into subject- predicate-language, and in that case the logics of both the types of universal statements would not diff :. The statement of the form 'Wherever .... there ....' can be translated in terms of 'Whatever is .... is ....' or 'All .... are ....'. In such a translation the metaphysical model of *ādhāra-ādheya*-relation is not abandoned, though the logical model of subject-predicate is explicity used. And Naiyāyikas would not object to such a translation. Of course the converse may not be true. The translation of the Buddhists' universal statement like 'Whatever is real is impermanent' into *ādhāra- ādheya*-language as 'Wherever there is reality there is impermanence' would not be acceptable to Buddhist logicians. It may be worth-noting in this context that Buddhist logicians use the language of *dharma* and *dharmin* which resembles the language of predicate and subject. The Nyāya language of *viśeṣya* and *viśeṣaṇa* (or *prakāra*) somewhat tallies with this language but the *ādhāra-ādheya*-language does not.

Of course this does not dissolve the problem of universals even at the hands of the Buddhist logicians. While stating the universal statement like 'Whatever is real is impermanent', the Buddhists avoid the language which may suggest that the common properties exist on par with the particulars; still they have to answer the question, as to what makes us attribute one and the same predicate to a number of particulars. At this stage the Buddhists have to consider positively the concept of universal. To say that universals do not exist on par with the ontological reals, is to give the negative account of universals, but to say that universals are the constructions of our mind (*vikalpa*) is to give the positive account. Buddhists thus accept universals as epistemic constructions. Here not only they differ from Naiyāyikas, but also suggest that in any act of inferrring our mental constructions play an important role. According to this view we construct the 'universal relation' of *vyāpti* on the basis of the particulars we observe. Naiyāyikas on the contrary would hold that we 'perceive' the universal relation, for universals for them are the perceptible ontological entities.

APPENDIX II

Two Concepts of Universality

In every post-Udyotakara theory of *anumāna* we find that the concept of *vyāpti* is predominent. '*Vyāpti*' is many times translated as the universal relation and the statement of *vyāpti* is generally supposed to be a universal statement. However, many a time it is ignored that the term universal is misleading. 'Universal relation' may mean a relation between two or more universals and a 'universal statement' may be taken to mean a statement which connotes such a relation between two or more universals. Thus the universal statement namely 'All men are mortal' may be interpreted as: "Humanity" (which is a common property of all men) is invariably related with "mortality" (which is a common property of all mortals). Similarly the *vyāpti*-statement like 'wherever there is smoke, there is fire' may be interpreted as 'Smokiness *(dhūmavattva)* is invariably related with fieriness *(vahnimattva)*'. This exhibits a realistic model of universality where universals are supposed to exist in the world as the 'common' properties shared by a number of things. *Vyāpti* thus conceived becomes a real relation amongst the real entities. Nyāya logicians seem to have accepted this model.

But the 'universal relation' need not necessarily presuppose the existence of universals. The universal relation may mean the particular relation with reference to all the concerned particular relata. Thus 'All men are mortal' in this sense may be interpreted as 'Every particular man is mortal'. The earlier meaning was 'the universal man' (humanity) is mortal', which should be distinguished from this meaning. According to this second concept of universality 'universal statement' is in fact a particular statement, but concerned with all particulars. This second model of universality seems to be accepted by Buddhist logicians.

It is worth noting in this context that the forms of *vyāpti*-statement accepted by Nyāya and Buddhist logicians differ from each other. Nyāya logicians are accustomed to use the form *'yatra yatra .... tatra tatra ....'. In this form of vyāpti* the *ādhāra- ādheya*-relationship is presupposed. The universal properties namely *hetu* and *sādhya* are

Both the assertions of the first CE namely A_1 and A_2 can be derived from the single statement Ha.

(B) As we have seen in § 3.3, the second CE is the conjunction of B_1 and B_2. An implication of B_1 and B_2 can be symbolised as

$(\exists x)$ Hx.Sx. $(x \neq a)$ .... (i)

Instead of such a proposition, Diṅnāga prefers to give a particular instance as a part of the statement of the second CE. For instance, if d stands for the subject of the positive instance, then the necessary qualifications that the instance should have can be symbolised as

Hd. Sd. $(d \neq a)$ .... (ii)

As we saw in § 3.8, this can be called the expression of 'Revised B_1'.

We find that (i) can be derived from (ii).

B_2 can be symbolised as

$(x)\ \{[Hx.\ (x \neq a)] \supset Sx\}$ .... (iii)

By conjoining (iii) and (ii) we get

$(x)\ \{[Hx.\ (x \neq a)] \supset Sx\}.\ Hd.\ Sd.\ (d \neq a)$.

We use this as the expression of the second CE.

(C) As we have seen in § 3.4, the third CE is the conjuction of C_1 and C_2. An implication of C_1 and C_2 can be symbolised as

$(\exists x)\ [\sim Sx.\ \sim Hx.\ (x \neq a)]$ .... (iv)

Here too, instead of such a proposition, Diṅnāga uses a particular instance as the part of the third CE. For instance, if d' stands for the subject of the negative instance, then the necessary qualifications that the negative instance should have can be symbolised as

$\sim Sd'.\ \sim Hd'.\ (d' \neq a)$ .... (v)

Following § 3.8 we call this an expression of 'Revised C_1'. Here too we find that (iv) can be derived from (v).

C_2 can be symbolised as

$(x)\ \{[\sim Sx.\ (x \neq a)] \supset \sim Hx\}$ .... (vi)

By conjoining (vi) and (v) we get

$(x)\ \{\ [\ \sim Sx.\ (x \neq a\ )]\supset \sim Hx\ ]\}\ .\ \sim Sd'.\ \sim Hd'.\ (d' \neq a).$

We use this as the expression of the third CE.

One who makes a *parärthānumāna*, first asserts the thesis and then for proving the thesis, asserts that the *hetu* which is used for proving the thesis, possesses all the three CEs. Therefore the total argument (i.e. the inferencial statement) will be of the form:

Thesis : Sa

First step: Ha

Second step: $(x)\ \{[\ Hx.\ (x \neq a)]\supset Sx\ \}\ .\ Hd.\ Sd.\ (d \neq a)$

Third step: $(x)\ \{\ [\ \sim Sx.\ (x \neq a)\ ]\supset \sim Hx\ \}\ .\ \sim Sd'.\ \sim Hd'.\ (d' \neq a).$

This is nothing but the symbolic expression of the form of *parārthānumāna* that we explained with illustration in § 3.8. We have called this form P_1.

(D) In § 3.9 we have seen that in *kevalānvayi* type of inference *vipakṣa* is non-existent. Hence such a *parārthānumāna* will lack the mention of any real negative instance.

The form of such *parārthānumāna* will be as follows:

Thesis : Sa

First step : Ha

Second step : $(x)\ \{\ [\ Hx.\ (x \neq a)\ ]\supset Sx\}.\ Hd.\ Sd.\ (d \neq a).$

Third step : $(x)\ \{\ [\ \sim Sx.\ (x \neq a)\ ]\supset \sim Hx\}$

This is nothing but the symbolic expression of the form of *parārthānumāna* that we explained with illustration in § 3.9. We have called this form P_2.

(E) It is clear both in P_1 and P_2 that the premises in these forms of *parārthānumāna* do not entail the thesis (i.e., the conclusion). Hence it is possible to accept the truth of the premises and yet to deny the conclusion. this is done in a way when an inference is accused of the fallacy called *viruddhāvyabhicāri*. There the opponent who points out the fallacy does not deny the truth of the premises presented by the disputant but he attempts to provide equally true premises which 'lead to' (again, not 'entail') the negation of the conclusion of the disputents argument.

The form of this fallacy which we discussed with illustration in §3.34 will be as follows:

(1) Sa

Because Ha

$(x) \{ [Hx . (x \neq a)] \supset Sx \} . Hd . Sd . (d \neq a)$

$(x) \{ [\sim Sx . (x \neq a)] \supset \sim Hx \} . \sim Sd' . \sim Hd' . (d' \neq a)$

But (2) $\sim$ Sa

Because Ga

$(x) \{ [Gx . (x \neq a)] \supset \sim Sx \} . Gf . \sim Sf . (f \neq a)$

$(x) \{ [Sx . (x \neq a)] \supset \sim Gx \} . Sf' . \sim Gf' . (f' \neq a)$

(3) Dharmakīrti's forms of *parārthānumāna*

In Chapter V we compared Dharmakīrti's notion of *parārthānumāna* with Diṅnāga's notion, and found that Dharmakīrti's notion differs from that of Diṅnāga in the following manner.

1. For Dharmakīrti both *sapakṣa-sattva* and *vipakṣāsattva* need not be stated as premises. The statement of both is neither logically necessary nor sufficient for proving the thesis. However, statement of *anvayavyāpti* or *vyatirekavyāpti* is necessary and becomes sufficient for proving the thesis when accompanied by *pakṣadharmatā*-statement.

2. *Anvayavyāpti*-statement and *vyatirekavyāpti*-statement entail each other.

3. Instanciation of *vyāpti* is not a necessary part of any premise of a *parārthanumana*.

4. The order of the premises of a *parārthānumāna* can be altered without any logical loss.

Thus Dharmakīrti would approve of the following forms of *parārthānumāna* (Please refer to § 5.12).

P_3 : Sa

Because Ha

$(x)(Hx \supset Sx). Hd. Sd. (d \neq a)$

P_4 : Sa

Because Ha

$(x)(\sim Sx \supset \sim Hx). \sim Sd'. \sim Hd'. (d' \neq a)$

P_5 : Sa

Because Ha

$(x)(Hx \supset Sx)$

P_6 : Sa

Because Ha

$(x)(\sim Sx \supset \sim Hx)$

P_7 : Sa

Because $(x)(Hx \supset Sx)$

Ha

P_8 : Sa

Because $(x)(\sim Sx \supset \sim Hx)$

Ha

We get P_5 by dropping the positive instanciation part in P_3. Similarly we get P_6 by dropping the negative instanciation part in P_4. We get P_7 and P_8 by changing the order of the premises in P_5 and P_6 respectively.

APPENDIX IV

I or O-propositions as conclusions.

While comparing Aristotalian formal fallacies such as illicit process of minor, undistributed middle ect. with similar formal fallacies noticed by Indian logicians, we have suggested that in Indian logic we cannot consider the validity or invalidity of the arguments, the conclusions of which are either 'particular affirmative' or 'particular negative' statements. (i.e. either I-statements or O-statements.) (See § 8.28 to § 8.31). In this note we are going to pose this itself as a problem and try to deal with it. The problem is like the following:

How can we make sense of the validity of the valid argument:

"All humans are mortal

Some Greeks are humans

Therefore some Greeks are mortal."?

Similarly how can we make sense of the invalidity of the invalid argument:

"Some humans are black

Some Greeks are humans

Therefore some Greeks are black."?

We may suggest that the second argument lacks *vyāpti*-statement and hence it is a case of *nyūna*. This is in tune with Aristotelian suggestion that in no valid syllogism both the premises are 'particular' statements. I.e., every syllogism must have at least one universal statement as its premise. (And both 'universal affirmative' and 'universal negative' statements can be treated as *vyāpti*-statements.) But the main problem is regarding the translation of 'particular' statements in the language of Indian logic. According to Indian logicians *pakṣa*-subject must stand for some given specific object. That object should not be of a 'general' character, but it is really a particular object. To use Russell's jargon, *pakṣa*-subject must be a proper name or a definite description. (*Pakṣa*-subject sometimes stands for a set of objects, rather than a single object. For instance the subject of the thesis 'sound is impermanent' namely 'sound' stands for the set of all sounds in the universe because there, sound is impermanent means every sound is impermanent.) However the predicate that we prove to be predicated of the *pakṣa*-subject is proved with its general character; the specific character of the object referred to by *sādhya* is not known. For example when a mountain is proved to be fiery on the basis of its being smoky, the *specific* mountain is proved to possess *some* fire. When sound is proved to be impermanent, it is not 'some' sound but 'the whole class of sounds' or 'every sound' that is proved to be impermanent. And sound is proved to be impermanent means sound is proved to belong to the class of impermanent objects, i.e. sound is proved to be *some* member of the class of impermanent objects. In such a case 'Some men are mortal' (i.e., some men are some members of the class of mortals) could never be the conclusion of any sound inference in Indian logic. However, it appears to me that there is a way in which such conclusions can be formally reduced to the statements which are fit for being the conclusions of some inferences in Indian logic.

Here we may direct our attention to the Nyāya inferences of the existence of God (*Īśvara*). When Nyāya philosophers advance cosmoteleological arguments for the existence of God, they really try to prove first that (some) God exists and then proceed to prove that God is one. When they try to prove that (some) God exists, what they try to prove is that '(some) designer of the world exists' or rather 'the world has some designer'. Thus the statement about (some) God turns out to be the statement about the world as a whole. By using the same technique, the statement that 'some men are mortal' can be translated into 'there are some mortal-men' i.e. into 'the world has some mortal-men'. This statement is about the given specific world whereas the predicate is not that specific.

Now let us come to the valid argument:

All humans are mortal

Some Greeks are human

Therefore some Greeks are mortal.

This argument can be restated as follows:

1. For all x and for all Y, if x has something having the properties Y-ness and human-ness, then x has something having the properties Y-ness and mortality.

2. The world possesses some things having the properties Greekness and humanness.

3. Therefore the world possesses some things having the properties Greekness and mortality.

This sort of re-statement becomes possible because

All humans are mortal means,

For all y if y is human then y is mortal.

From this it follows that

For all x and for all Y if x is a place in which there are Y's which are human, then x is a place in which there are Y's which are mortal.

In other words

For all x and for all Y if x has something having the properties Y-ness and humanness then it has the properties Y-ness and mortality.

It is true that in this kind of version of *vyāpti*-statement, we have more variables than one, out of which one is subject-variable and others are predicate- variables. But such universal statements are not unfamiliar to Indian logicians.

Thus the above argument can be stated in Sanskrit as:

1) यत्र यत्र यत्प्रकारविशिष्टमनुष्यवत्त्वं तत्र तत्र तत्प्रकारविशिष्ट-मर्त्यवत्त्वम् ।

2) इदं जगत् ग्रीकत्वविशिष्टमनुष्यवत् ।

3) तस्माद् इदं जगत् ग्रीकत्वविशिष्टमर्त्यवत् ।

Although this method is artificial, it is formally acceptable. It shows that formally speaking Indian logic is as complete as Aristotelian Syllogistic. I.e., it can recognise all the arguments as valid or otherwise which Aristotle in his Syllogistic recognises as valid or otherwise (respectively) and the other way about.

This method also shows that Aristotle's scheme of major, minor and middle term is different from Indian logicians' scheme of *sādhya*, *pakṣa* (i.e. *pakṣa*-subject) and *hetu*. In the above argument what we took to be minor term (namely 'Greek') in terms of Aristotelian Syllogism, after translation in the form of Indian (rather Dharmakīrti's) *parārthānumāna*, turned out to be a part of *sādhya* (which is said to be parallel to 'major term' in Aristotelian Syllogism.). And some new term ('this world') was introduced in the place of 'minor term' (which is said to be parallel to *pakṣa*-subject.)

If what I have said so far holds water, then this suggests that Indian logicians' scheme (especially Dharmakīrti's scheme) has an advantage over Aristotle's Syllogistic. The conclusion of any syllogism can be reduced to *pakṣa* (thesis) of some *parārthānumāna* and then we can try to reduce the form of syllogism to the form of *parārthānumāna*. In Indian logic we have basically only two forms of valid *parārthānumāna*, modus ponuns and modus tollens (i.e. *sādharmyavat prayoga* and *vaidharmyavat prayoga* in Dharmakirti's language, see § 5.12). We have seen that any valid syllogism can be reduced to either of the two forms. This gives us a simpler and more systematic method of deciding whether the given syllogism is valid or not. We do not have to keep in mind and apply so many valid forms of syllogism, doing which is more cumbersome and less systematic.

Indian method of *parārthānumāna* is more appealing philosophically also. It shows that while inferring, something (*pakṣa*-subject) must be given to us with logical definiteness. We infer it to be related with something (*sādhya*) on the basis of something else (*hetu*) which relates itself with both. While inferring we are concerned with *pakṣa*, *hetu* and *sādhya* which have ontological existence and which represent a certain kind of relational triad. We are not just concerned with three *terms*, which we can relate with each other in varied ways that cannot be easily systematised or simplified.

Bibliography

1. —; *Bṛhadāraṇyakopaniṣad*, Ānandāśrama Sanskrit Granthāvali, No. 64, Pune, 1911.
2. Agniveśa; *Carakasaṁhitā* (First Part) with Hindi Commentary by SHASTRI, Kashinath and CHATURVEDI, Gorakhnath; Chaukhambha Sanskrit Saṁsthana, Varanasi, Fourth ed; 1976.
3. Agniveśa; *Carakasaṁhitā* with Āyurveda-dīpikā of Cakrapāṇidatta; Nirnaya Sagara Press, Bombay, 1922.
4. Agniveśa; *Carakasaṁhitā* Vol. II, Ed. Jadhavaji Trikamji Acharya, Third Ed. Bombay, 1941.
5. Akalaṅka; *Laghīyastrayādisaṅgraha*, Manikchand Digambar Jain Granthamala Samiti, Bombay, 1916.
6. Arcaṭa; *Hetubinduṭīkā*, Gaekwad's Oriental Series No. CXIII, Baroda, 1949.
7. ATHALYE and BODAS; *Tarkasaṅgraha of Annambhaṭṭa* Bombay Sanskrit Series No. LV, Bhandarkar Oriental Research Institute, Poona, 1974.
8. BANDYOPADHYAY, Nandita; *The Concept of Logical Fallacies* Sanskrit Pustak Bhandar, Calcutta, 1977.
9. BAPAT, Shailaja S.; The Term *Arayavin* and *Aṁśin* in Indian Philosophy, Proceedings of the All-India Oriental Conference, 28th Session, Dharwad, Nov. 1976.
10. BARLINGAY, S.S.; *A Modern Introduction to Indian Logic*, National Publishing House, New Delhi, Second Edition, 1976.
11. BARLINGAY, S.S.; 'Distinguishables and Separables', Indian Philosophical Quarterly, Vol. II, No. 2, Jan. 1975.
12. BARLINGAY, S.S.; *Bhāratīya Tarkaśāstra kī Rūparekhā*, (Hindi), Rajasthan Hindi Grantha Academy, Jaipur, 1976.
13. BARLINGAY, S.S.; 'Wholes : Mechanical and Organic' included in BARLINGAY S.S.; *Beliefs, Reasons and Reflections*, Indian Philosophical Quarterly Publication, Pune, 1983.
14. Bhartṛhari; *Vākyapadīya (Brahmakāṇḍa)*, Kashi Sanskrit Series 124, Chaukhamba Sanskrit Sansthan, Varanasi, Third Edition, 1975.

15. Bhāsarvajña; *Nyāyabhūṣaṇam*, Ṣaḍdarśana Prakāśana Pratiṣṭhāna, Varanasi, 1968.

16. CHATTERJEE, Satischandra; *The Nyāya Theory of Knowledge*, University of Calcutta, Second Ed. 1950.

17. COHEN, Morris R. and NAGEL Ernest; *An Introduction to Logic and Scientific Method*, Allied Publishers Limited, New Delhi; Third Indian Reprint, 1976.

18. DEVENDRA MUNI SHASTRI; *Jaina Darśana Svarūpa aura Viśleṣaṇa* (Hindi), Shri Taraka Guru Jaina Granthamala, Udaypur, 1975.

19. Dharmakīrti; *Pramāṇavārtika*, Buddha Bharati, Varanasi, 1968.

20. Dharmakīrti; *Vādanyāyaprakaraṇa*, critically edited by Swami Dwarikadas Shastri, Buddha Bharati, Varanasi, 1972.

21. Dharmottara; *Nyāyabinduṭīkā*, Edited with Hindi translation and notes by Shrinivasa Shastri, Sahitya Bhandar, Merath, 1975.

22. Diṅnāga; *Nyāyamukha*, Edited in Chinese and Tibetan and translated by TUCCI, Guiseppe; Materialen Zur Kunde des Buddhismus.

23. Diṅnāga (Śaṅkarasvāmin ?); *The Nyāyapraveśa* Part I, Critically Ed. with notes and introduction by DHRUVA A.B.; Oriental Institute, Baroda, 1968.

24. EATON, Ralph M; *General Logic*, Charles Scribner's Sons, New York, 1959.

25. GALLIE, W.B.; 'The Function of Philosophical Aesthetics' *Aesthetics and Language*, Ed. William Elton, Basil Blackwell, Oxford, 1970.

26. Gaṅgeśopādhyāya; *Sāmānyanirukti*, Mithila Institute of Post-Graduate Studies and Research in Sanskrit Learning, Darbhanga, 1970.

27. Gauḍapāda; *Gauḍapādakārikā*, Ed. with trans. and notes by R.D. Karmarkar; Bhandarkar Oriental Research Institute, Poona, 1953.

28. Gautama (Akṣapāda); *Nyāyasūtrāṇi*, with *Bhāṣya* of Vātsyāyana and *Vṛtti* of Viśvanātha, Anandashram Sanskrit Granthavali No. 91, Pune, 1922.

29. GOKHALE, P.P.; 'The Philosophical Position of Jayarāśi-bhaṭṭa', Indian Philosophical Quarterly, Vol. V, No. III, April 1978.

30. HAMBLIN, C.L.; *Fallacies*, Methuen and Co. Ltd., London, 1970.

31. Jayantabhaṭṭa; *Nyāyamañjarī*, Vols. I and II, Kashi Sanskrit Series, The Chawkhamba Sanskrit Series Office, Part I, Second Ed. 1971, Part II 1934.

32. Jayarāśibhaṭṭa; *Tattvopaplavasiṁha*, Ed. with introduction by SANGHAVI and PARIKH, Gaekwad's Oriental Series No. LXXXVII, Baroda, 1940.

33. Kaṇāda; *Vaiśeṣikasūtra*, Critically Ed. by Muni Shri Jambuvijayaji; Gaekwad's Oriental Series No. 137, Baroda, 1961.

34. KATSURA, Shoryu; 'Sanskrit Fragments of the *Pramāṇa-samuccaya*' Journal of Indian Philosophy, Vol. 3 Nos. 1/2, September-December 1974.

35. KEITH, Arthur Berriedale; *Indian Logic and Atomism*, Humphrey Milford; Oxford University Press, 1930.

36. KEYNES, John Maynard; *A Treatise on Probability*, Macmillan and Co. Ltd., London, 1948.

37. Mādhava, Sāyaṇa; *Sarvadarśanasaṅgraha*, The Bhandarkar Oriental Research Institute, Poona, 1924.

38. MATTLAL, B.K.; 'A Note on Nyāya Fallacy Sādhyasama and Petitio principii', Journal of Indian Philosophy, Vol. 2 Nos. 3-4, March-June, 1974.

39. MILL, J.S.; *A System of Logic, Ratiocinative and Inductive*, Collected Works of John Stuart Mill, Vol. VII, University of Toronto Press, Toronto and Buffalo, 1973.

40. MOORE, G.E.; 'A Defence of Common Sense' included in *Philosophical Papers* by Moore, G.E., George Allen and Unwin Ltd., London, 1959.

41. MULLATTI, L.C.; *The Navya-Nyāya Theory of Inference*, Karnatak University, Dharwad, 1977.

42. Nāgārjuna; *Madhyamakaśāstra*, Buddhist Sanskrit Texts No. 10, The Mithila Institute, Darbhanga, 1960.

43. NATHMAL, Muni; *Jaina Darśana Manana Aura Mīmāṁsā*, (Hindi) Adarsha Sahitya Sangha, Churu, 1973.

44. Patañjali; *Vyākaraṇamahābhāṣyam (Navāhnikam)*, Kashi Sanskrit Granthamala, No. 153, Chaukhamba Sanskrit Series, Benares, 1954.

45. POTTER, Karl H. (Ed.); *Encyclopedia of Indian Philosophies* Vol. II, Motilal Benarsidass, First Ed., Delhi, 1977.

46. Praśastapāda; *Praśastapādabhāṣyam*, Kashi Sanskrit Series, Chaukhamba Sanskrit Pustakalaya, Kashi, 1923.

47. RANDLE, H.N.; *Indian Logic in the Early Schools*, Humphrey Milford, Oxford University Press, 1930.

48. RYLE, Gilbert, 'Systematically Misleading Expressions', Proceedings of Aristotalian Society, Vol. XXXII, 1931-32, pp. 139-70.

49. Samantabhadra; *Āptamīmāṁsā Pramāṇaparīkṣā Ca*, Bharatiya Jaina Siddhanta Prakashini Samstha, Benares, 1914.

50. Śaṅkarācārya; *Brahmasūtrabhāṣyam*, Shri Kamakoti-koshasthanam, Madras, 1954.

51. Siddhasenadivākara; *Sammatitattvasopānam*, Shri Labdhasurishvara Jaina Granthamala, Chhani, Baroda, 1946.

52. STCHERBATSKY, F. Th.; *Buddhist Logic* Vol. I, Dover Publications, New York, 1962.

53. SUGIURA, Sadajiro; *Hindu Logic as preserved in China and Japan*, University of Pennsylvania, Series in Philosophy No. 4, Philadelphia, 1900.

54. TUCCI, Giusepee (Ed.); *Pre-Diṅnāga Buddhist Texts on Logic from Chinese Sources*, Gaekwad's Oriental Series Vol. XLIX, Baroda, 1929.

55. Udyotakara; *Nyāyavārtika*, Ed. by Laxmanshastri Dravid, H.D. Gupta and Sons, Benares, 1915.

56. UI, H; *The Vaiśesika Philosophy*, Chowkhamba Sanskrit Series Vol. XXII, Varanasi, 1962.

57. Umāsvāti; *Tattvārthasūtra*, Commented by SANGHAVI, Sukhalalji, Jain Sanskrit Samshodan Mandal, Benaras, Second Ed. 1952.

58. Vācaspatimiśra; *Bhāmatī*, Bibliotheca Indica, The Asiatic Society of Bengal, Benares, 1880.

59. Vācaspatimiśra; *Nyāyavārtikatātparyaṭīkā*, Vols. I and II, Ed. by Rajeshwarashastri Dravid; Haridas Sanskrit Granthamala No. 24, Chaukhamba Sanskrit Series, Benares, 1925.

60. Vādidevasūri; *Pramāṇanayatattvālokālaṅkāra*, rendered into English with a commentary by Dr. Bhattacharya, Hari Satya, Jaina Sahitya Vikas Mandal, Bombay 1967.

61. VIDYABHUSANA, Satis Chandra; *A History of Indian Logic*, Motilal Banarsidass, Delhi, First Edition reprinted, 1978.

Index of Terms

Index of Names